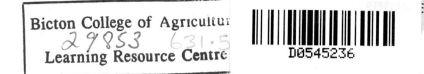

The RHS Encyclopedia of Practical Gardening

PLANT PROPAGATION

PHILIP McMILLAN BROWSE

Editor-in-chief Christopher Brickell
Technical Editor Kenneth A. Beckett

Philip McMillan Browse is a partner in a horticultural consultancy in Truro, Cornwall, and a former President of the International Plant Propagators' Society (1978).

MITCHELL BEAZLEY

The Royal Horticultural Society's Encyclopedia of Practical Gardening © Octopus Publishing Group Ltd 1979, 1992, 1999

The Royal Horticultural Society's Encyclopedia of Practical Gardening: Plant Propagation © Octopus Publishing Group Ltd 1979, 1992, 1999

First published in 1979
Second edition 1992
Reprinted 1993 (twice), 1994, 1995, 1996, 1997, 1998
New edition 1999 Reprinted 2000, 2002

ISBN 1 84000156 9

Edited and designed by Mitchell B   of
Octopus Publishing Group Ltd
2-4 Heron Quays, London E14 4JP
Produced by Toppan Printing Co (HK) Ltd.
Printed and bound in China

Contents

Introduction 3–5
Glossary 6–9

TOOLS AND EQUIPMENT 10
Environmental control 12–13
Knives 14–15
Secateurs 16–17
Containers 18–21
Composts 22–23
Rooting hormones/Wounding 24–25
Watering 26–27
Fertilizers 28–29
Hygiene 30–31
Pests 32–33
Diseases 34–35

SEEDS 36
Collecting and storing 38–39
Sowing in containers 40–41
The developing seed 42–45
Alpines 46–47
Bedding plants 48–49
Herbaceous plants 50–51
Ferns 52–53

Trees and shrubs 54–57
Buying and collecting 58–59
Extracting 60–61
Storing/Breaking dormancy 62–65
Preparing a seedbed 66–67
Sowing in a seedbed 68–69
Exotic trees and shrubs 70–71

ROOTS 72
Root cuttings 74–79
Tuberous roots 80–81

MODIFIED STEMS 82
Tubers 84–85
Rhizomes 86–87
Corms 88–89
Bulbs 90–91
Bulblets and bulbils 92–93

Bulb scaling 94–95
Scooping and scoring bulbs 96–97
Division 98–101
Offsets/Runners 102–103

STEMS 104
Layering/Simple layering 106–109
Air layering 110–111
Tip layering 112–113
Stooling 114–115
French layering 116–117
Dropping 118–119

Stem cuttings 120
Making a stem cutting 122–125
Soft woods 126–129
Green woods 130–131
Semi-ripe woods 132–133
Evergreens 134–137
Hard woods 138–145
Conifers 146–149
Sub-shrubs 150–151
Rhododendrons 152–153
Heathers 154–155

LEAVES 156
Leaf-petiole cuttings 158–159
Midrib cuttings/Lateral vein cuttings 160–161
Leaf slashing 162–163
Leaf squares 164–165
Monocot leaves 166–167
Foliar embryos 168–169

GRAFTING/Whip-and-tongue 170–173
Apical-wedge 174–175
Side-wedge 176–177
Side-veneer 178–179
Shield-budding 180–181
Rose budding 182–183
Chip-budding 184–185

Index 186–191
Acknowledgements 191

Introduction 1

One of the most exciting projects that a gardener can undertake is to propagate his own plants. Few experiences can match the thrill and awe felt by a gardener who has successfully germinated an even crop of bedding plant seedlings or a particularly difficult woody plant seed, or who has managed to root a cutting or unite two grafted parts of a plant.

Plant propagation has been practised ever since early man abandoned his itinerant life and settled down on the land, where he started to grow his own food. The techniques of vegetative propagation developed in such ancient civilizations as those of the Babylonians and the Chinese are still relevant and in use in the twentieth century.

Using this book

The object of this book is to try to present some of the basic facts on which successful plant propagation is founded, so that the gardener wishing to venture into this field will find the elementary principles explained and will then only have to learn their application from experience.

It is not intended that this book will be a complete answer – it can only be a guide. Inevitably, some techniques of plant propagation have not been included either because they are too complicated or because they do not always produce satisfactory results. Vegetables have not been included as they are given detailed coverage in a companion volume, *Vegetables*. However, a comprehensive range of techniques is described so that the gardener can propagate most plants, either from seed or vegetatively.

The gardener can discover the way to propagate more than 700 genera of plants by referring to the index at the back of the book, where the appropriate method is listed. A few plants, such as rhododendrons, have been dealt with in detail, and so have a page to themselves, because there are several equally satisfactory ways to propagate them.

How this book is arranged

The concept and approach in this book is entirely my own, based on my experience gained over the previous 20 years evolving different systems of plant propagation and teaching both horticultural students and amateur gardeners.

My aim has been to try to present propagation techniques in their logical sequence, in seven separate sections. For example, I have placed layering in the same section as stem cuttings as they are both ways of inducing roots on a stem. Until now, plant propagation has often been presented in a confusing manner: root, stem and leaf cuttings have frequently been lumped together, when logically there is little or no relationship between them.

The book starts with a glossary of propagation terms that the gardener may not know. Although the main text is written in a non-technical style, some technical terms have inevitably been included, and it is to the glossary that the gardener should refer if he is puzzled by the use of a certain word.

The section on tools and equipment is essential reading before the gardener sets out to do any propagating. Here he will find explained the importance of cold frames, propagators, mist units, etc., and he will discover tips on how to select suitable pots and seed trays from the bewildering array that is available. The correct use of composts, fertilizers and rooting hormones is also discussed, and he will learn how to tackle the ubiquitous problems of pests and diseases.

The section on propagating from seeds is divided into two parts: seeds in general, and tree and shrub seeds, where the problems of dormancy are explained. The information on how to collect, store and germinate various kinds of seeds and how to look after seedlings is of vital importance to any gardener.

The book then presents a wide range of ways to propagate vegetatively. It starts with a section on roots and progresses through to

Introduction 2

sections on modified stems, stems and leaves. The last section gives full details on grafting and budding techniques, and there is a comprehensive index at the back of the book.

Dispelling the mystique

The practice of plant propagation is inclined to be surrounded by a certain mystique, despite the enlightened modern tendency to share knowledge. The "art" of plant propagation and the possession of "green" fingers are not magical powers confined to a few fortunate gardeners; plant propagation is an entirely rational and logical technique, which, if practised with knowledge and understanding, can be relatively simple and easy. Undoubtedly some people do seem capable of conjuring success without any basic knowledge and of carrying out the various operations instinctively, but these people have, perhaps by subconscious observation, noted correct conditions and timing. The gardener who is gifted with "green" fingers is not dissimilar to the person who is "good" with animals or who has the capability to lead other people. It is simply the result of an ability to observe particular conditions.

Nothing, however, can supersede the value of real knowledge and understanding that the gardener needs when propagating a plant. Only once he is fully versed in the basic propagating techniques can he hope to be really successful.

Although plant propagation is completely logical and explainable, some gardeners are definitely more successful than others; and it is here that the "art" of plant propagation cannot be discounted. The "art" is in interpreting information: it is possible to define the state of plant materials, the condition of plants or the effect of a particular environment, but success derives from being able to transfer this knowledge into practice and interpret these aspects in relation to a situation that is continually, but often almost imperceptibly, changing.

The many phases of plant propagation

The actual process of propagation is only one of many phases in the production of a plant.

The other phases are selecting suitable plant material; preparing it so it has a high capacity to regenerate; then providing suitable conditions in which the plant material can regenerate; and ensuring its survival until the final phase of establishing the plant material as an integrated, self-supporting new plant.

Selecting suitable plant material

Often the most overlooked, but one of the most significant, phases is the consideration and choice of suitable material from which to propagate. It is well worth the extra time and effort to assess the available plant material critically so that the best selection is chosen, and new plants are not produced from inferior stock. Only the best forms and selections of a plant should be earmarked for propagation, and they must always be from healthy stock, free from virus infections. Many plants, especially the older and popular selections, will have deviated from the normal to some extent. Despite their varietal name, they may differ quite considerably and will exist in several clones, so bear this in mind when choosing plant material for propagation.

Another limitation that should be considered if propagating by vegetative methods is that the capacity of the plant to regenerate will be affected by the age of the cutting and its parent plant, as well as the age of the variety from which it is taken.

Plant material of the current year's growth will regenerate more readily than older material, and the highest rooting response will be found in a plant that is juvenile, i.e. immature and unable to produce flowers or fruit. As soon as a seed germinates and produces a juvenile plant, it begins to "age" and its capacity to regenerate starts to decline. Most plants subsequently enter a mature phase when their regenerative abilities continue to decline.

Old plants and older varieties will exhibit very low levels of response. Pruning or forcing a plant will only recover a little of this capacity to regenerate. Thus the gardener must be prepared to accept that old plants and varieties will be difficult to propagate. For example, a deciduous azalea, of the Exbury type, which was germinated from seed only about 60 years ago, will be much easier to propagate than a Ghent azalea, which would have been originated over 160 years ago. It is important to realize this distinction, and that all plants derived from one selected form must, physiologically, be the same age, regardless of when they were vegetatively propagated.

Preparing the plant material

Having chosen the most desirable forms, the next phase is to prepare the material so that, when the time comes for propagation, it will possess its maximum capacity to regenerate. Preparation of highly regenerative material can be done by growing techniques such as pruning, feeding and watering, or by forcing the plant in a warmer environment than normal.

Providing suitable conditions for regeneration

Next, it is necessary to stimulate this plant material to regenerate as a new plant by encouraging the processes that cause the development of a new and complete plant. This can be done by placing the plant material in a suitably controlled environment, such as a cold frame or propagator. This will not only speed up regeneration but also lessen the chances of the plant material dying from rotting, disease or exhausted food reserves. A stem can also be encouraged to produce roots by dipping it in a rooting hormone or wounding it towards its base.

The ability of plant material to regenerate is also influenced by the different seasons.

Always propagate a plant during the season recommended by this book.

Ensuring its survival

Once the plant material is in its propagating environment, it is vital to ensure its survival until it becomes established.

The only way to do this is by maintaining absolute hygiene in the propagating environment and by thoroughly cleaning all tools and equipment. Also, treat the plant material with a systemic or copper-based fungicide, and protect it with a general or specific pesticide. The shorter this survival period the less time there is for things to go wrong.

Establishing the new plant

As soon as the plant material has regenerated, the last phase in successful propagation is the establishment of this young material as an integrated, self-supporting new plant. When any cutting, such as a leaf or stem, is taken, the new parts that are required to form a complete new plant will need time to become fully integrated with the original cutting. Just because a stem cutting produces roots does not imply a new plant – both systems must grow sympathetically so that a balanced and integrated growth is achieved. It is often relatively simple to persuade a cutting to regenerate a missing part, but it is more difficult to establish the plant material. This has to be done by weaning it from its protected environment and hardening it off until it is a self-supporting individual that can grow happily in a normal environment, whether this is indoors or outdoors. This is often the hardest part of plant propagation.

The path of success

Provided the gardener uses this book to understand the principles and basic practical tasks of plant propagation and then follows his judgement in relation to a particular plant, he can always approach propagation with the confidence that underlies success.

Glossary 1

Adventitious Arising casually or in unusual places. A bud or a root, for example, that develops from tissue that would not normally give rise to such an organ.

Aeration The level to which air may freely penetrate or permeate a dense medium, for example, compost.

Air layering see page 110-11.

Apical bud The active bud on the end of a shoot.

Axil The upper angle between a leaf, or leaf-stalk, and the stem from which it grows.

Axillary bud The bud situated in the axil.

Bare rootstock Rootstock lifted from the open ground to be used for bench grafting.

Basal At the base or bottom. The basal cut on a cutting or scion is the cut made at the bottom. The basal shoot or root on a plant is the bottom one.

Basal or base plate The flattened or squat conical stem within a bulb.

Bench grafting Grafting on to a rootstock that is movable, that is, it is pot-grown or bare rootstock. The grafting operation can thus be carried out on a bench.

Blanching The exclusion of light from a stem so that its green colouring (chlorophyll) disappears. Blanching often causes rapid stem growth.

Block up the lights To raise the lid of a cold frame by propping it up with blocks of wood or similar material.

Bottom heat The warmth, normally provided artificially, from under the compost in, for example, a propagator, to encourage the initiation and development of roots.

Broadcast sowing The uniform and even distribution of seeds all over a seedbed (as opposed to sowing in drills).

Bud-break The end of the dormant season, when stem buds are induced to grow, usually when temperatures rise above 5°C/41°F.

Budding The process of grafting a bud on to rootstock.

Bud grafting An alternative term for budding.

Bud-stick The selected and prepared stem from which the buds are taken for budding.

Bud-trace The "inside" (back) of a bud attached to the woody part of a stem.

Bulb see page 90.

Bulbil see page 91.

Bulblet see page 92.

Bushel An Imperial measure of volume equal to 1.28 cu ft; for example, the amount of compost that will fit into a box 22 in × 10 in × 10 in without compacting.

Callus The protective wound tissue developed by a plant on any damaged surface.

Cambium The simple basic cell making up the actively growing tissues of a stem, root or leaf from which the various conducting tissues develop.

Capillarity The process by which water rises above its normal level through a series of very small spaces, for example, through sand. The smaller the spaces the higher the water rises.

Chinese layering An alternative term for air layering.

Chip-budding see page 184.

Church window effect An exposed cut surface on a rootstock or scion once they have been grafted together.

Clone A group of genetically identical plants produced vegetatively or asexually from a single parent plant.

Closed case see page 13.

Cold frame see page 12.

Compatibility A rootstock and a scion that are able to knit together to make a new plant, that is they will make a successful graft union. Incompatibility may be caused by uncongenial chemical reactions between rootstock and scion or by too great a diversity in the juxtaposition of their plant tissues. Delayed incompatibility refers to an apparently successful graft union that subsequently breaks because there has been no close bonding of tissues between rootstock and scion.

Compost see page 22.

Compound layering A technique similar to simple layering in which the stem is layered several times along its length. This technique, which is rarely practised, is also known as serpentine layering.

Continuous layering An alternative term for French layering.

Corm see page 88.

Cormel see page 88.

Crenation The indentation on a leaf that has round toothed edges.

Crown The part of a plant at or above ground level that normally produces stems. Herbaceous plants are usually referred to as having crowns; so, occasionally, are woody shrubs that tend to grow densely.

Crown bud A bud on the crown of, for example, an herbaceous plant which rests during the dormant season and will then develop in the subsequent growing season.

Cryptic fruited plant A plant in which the seed or seeds are concealed in the fruit by extensive wing or flesh.

Cutting A separated piece of root, stem or leaf that has been prepared solely to propagate a new plant.

Delayed compatibility *see* Compatibility.

Desiccate To lose water, to dry out; to wilt.

Dibber A tool used to make a hole in which to plant a cutting, seedling or small plant. Dibbers are available in various sizes that are related to the required size of the hole.

Dibble in To use a dibber.

Dicotyledon A flowering plant that produces two seed leaves at germination.

Differentiate The process of changing from one or more simple plant cells, for example in the cambium, into a particular specialized organ or tissue, for example roots or conducting tissues.

Dominant The effect of one plant organ on another, usually with regard to their position. For example, on a stem there is apical dominance over other lower buds.

Dormant Asleep. A dormant seed or plant is one that is in a temporary resting state while it survives adverse climatic conditions. *See* page 62.

Dropping *see* page 118.

Embryo The basic component of a new plant in a seed that, given the right conditions, will germinate and develop into a new plant.

Endosperm A tissue within the seed, normally used for food storage.

Ericaceous Belonging to the rhododendron and heather family. Ericaceous plants require acid soil conditions for normal growth.

Etiolation The blanching of a stem by the exclusion of light without first exposing it to light.

Eye A bud.

F$_1$ seed *see* page 39.

F$_2$ seed *see* page 39.

Feathered growth A regular and uniform series of lateral branches that are usually fairly short.

Field grafting Grafting on to a rootstock that is established and growing in normal soil in the open ground.

Foliar Of the leaf.

Foliar embryo *see* page 168.

French layering *see* page 116.

Fungicide A chemical that will kill fungi.

Genetics *see* page 39.

Germination *see* page 42.

Girdling *see* page 107.

Grafting *see* page 170.

Greenwood cutting *see* page 130.

Harden off *see* page 44.

Hardwood cutting *see* page 138.

Heading-back The removal of all rootstock above a graft union.

Heel cutting *see* page 124.

Heel in To store plant material, normally in the open ground. Stems, cuttings or plants are placed together in an upright or inclined position in a trench, which is then filled with soil and firmed.

Herbicide A chemical that will kill plants; normally refers to weedkillers.

Humidity The amount of water vapour in the atmosphere. Relative humidity is the amount of water in the atmosphere, relative to it being saturated, at a particular temperature. (Warm air will hold more water than cool air.)

Husbandry The understanding and care of plants.

Hybrid A plant produced by cross-pollinating two or more species.

Hypocotyl That part of a seedling which is between the shoot system and the root system. The area in which root tissues change to stem tissues, or vice versa.

Imbibe To drink in; absorb water.

Incompatibility *see* Compatibility.

Inhibit The suppression of a particular growth or developmental pattern.

Initial The very first development of a new plant organ, for example a root, when it is just possible to distinguish it amongst undifferentiated cells.

Insecticide *see* Pesticide.

Internode The stem between two adjacent nodes.

Irishman's cutting *see* page 101.

Juvenile A young, immature plant before it reaches sexual maturity and, hence, before it reaches its ability to flower.

Glossary 2

Latent bud A bud that has been produced normally, which for various reasons will not develop further unless stimulated by some unusual circumstance.

Lateral On the side. A lateral bud or root is on the side of a stem or root as distinct from being at the top or the base.

Layer A stem that is in the process of being layered. Also applied to the rooted plant at the time of and just after separating from the parent.

Leader shoot The shoot that is dominating growth in a stem system, and is usually uppermost.

Leaf-bud cutting see page 122.

Leaf cutting see page 156.

Leaf-fall The season when a leaf has been isolated from its stem by the development of a corky abscission layer.

Leaf-petiole cutting see page 158.

Leaf slashing see page 162.

Leaf square see page 164.

Leg A single stem that is developed to raise the main head of a tree or shrub.

Light The glass or plastic, normally enclosed by wood or metal, used for covering a cold frame or closed case.

Line out To plant out young plants or cuttings fairly close together in a straight line in the open ground.

Liner An established newly propagated plant that is ready for planting out or growing on in a nursery bed.

Long Tom A pot about half as deep again as a normal pot.

Mallet cutting see page 125.

Marcottage see Air layering.

Mature A plant that can produce flowers and, hence, reproduce sexually.

Midrib cutting see page 160.

Mist unit see page 13.

Modified stem see page 82.

Monocot leaf see page 166.

Monocotyledon A flowering plant that produces only one seed leaf.

Mound layering An alternative term for stooling.

Mycorrhiza The beneficial association between some fungi and the roots of some plants, in which the fungi fulfil many of the functions of the plant's root system and in return receive carbohydrates as food.

Node The place where a leaf joins the plant's stem and subtends an axillary bud.

Nurse graft A rootstock that supports a scion until it roots itself. A grafting technique often used in the production of, for example, clematis, cotoneaster and wisteria.

Offset see page 102.

Peat pellet see page 21.

Perennation The survival of vegetative plant parts during the dormant season.

Pericarp The wall of a fruit.

Pesticide A chemical used to kill a pest; a general term to cover insecticide, vermicide, acaricide, etc.

Petiole The stalk of a leaf.

Photosynthesis The process by which a green plant is able to make carbohydrates from water and carbon dioxide, using light as an energy source and chlorophyll as the catalyst.

Plumule The primitive shoot in an embryo.

Polarity The state of having two opposite poles, that is a top and a bottom.

Polythene tunnel see page 13.

Presser board A piece of flat wood with a handle used to firm and level compost.

Prick out see pages 43–4.

Propagator see page 13.

Prothallus The sexual generation of a fern.

Radicle The primitive root in an embryo, later becoming the first seedling root.

Reaction The degree of acidity or alkalinity in some soil or compost. Reaction is measured on the pH scale, on which pH 7.0 is neutral; lower figures indicate increasing acidity, whereas higher figures indicate increasing alkalinity.

Regeneration The development of any missing parts on propagated material (for example roots on a stem cutting) to make up a complete plant.

Relative humidity see Humidity.

Respiration The process by which a plant liberates energy for its growth processes.

Rhizome see page 86.

Ripe wood Almost hard wood.

Root cutting see page 74.

Rooting hormone see page 24.

Rose budding see page 182.

Runner see page 102.

Scale leaf A leaf modified in the form of a scale, often occurring underground on a modified stem.

Scaling a bulb *see* page 94.
Scarification *see* page 63.
Scion The portion (a single bud or stem) grafted on to a rootstock. Once established, it becomes the major part of the stem system on the new plant.
Scooping *see* page 96.
Scoring *see* page 96.
Seedling rootstock Rootstock that has been produced from seed as opposed to rootstock that has been propagated vegetatively.
Seed lot A collection of seeds from a particular plant or plants.
Semi-hard wood An alternative term for semi-ripe wood.
Semi-ripe wood cutting *see* page 132.
Serpentine layering An alternative term for compound layering.
Sessile Stalkless. A sessile leaf has its midrib and leaf blade attached directly to a plant's stem at a node.
Shield-budding *see* page 180.
Side-veneer grafting *see* page 178.
Side-wedge grafting *see* page 176.
Simple layering *see* page 106.
Softwood cutting *see* page 126.
Soil block *see* page 21.
Spit depth The depth of the blade on a normal digging spade; about 10 in.
Splice graft An alternative term for whip grafting.
Station sowing The individual sowing of seeds at a predetermined spacing in the site in which they will grow until pricking out or harvesting.
Stem cutting *see* page 120.
Stock An alternative term for rootstock.
Stolon A general term that is often used to cover a wide range of modified stems or parts of modified stems. Because of its confused usage, it has been ignored in this book as a definitive term.
Stooling *see* page 114.
Stratification *see* page 64.
Strike off To remove excess compost above the rim of a pot or seed tray, using a presser board, a piece of wood, edge of the palm or fingers.
Sub-shrub cutting *see* page 150.
Sub-terminal shoot A shoot immediately behind a leader shoot that usually grows actively but not quite as vigorously as a leader shoot.

Succulent A condition in certain plants that has developed as a response to a lack of readily available fresh water. A succulent plant is capable of storing relatively large quantities of water.
Sucker A shoot growing either from a stem or a root at or about ground level.
Systemic Capable of permeating the whole of a plant.
T-budding An alternative term for shield-budding.
Terminal bud The bud that terminates growth at the top of a stem and remains in a resting stage during the dormant period.
Tip layering *see* page 112.
Top working The grafting of a rootstock at standard or half-standard height.
Transpiration The process by which a plant naturally loses water, mainly through its leaves.
Trench layering An alternative term for etiolation layering.
Tuber *see* page 84.
Tuberous root *see* page 80.
Turgid Plant material that contains its full complement of water and is not therefore under stress.
Union The join where a rootstock and scion have been grafted so they produce common and continued growth.
Vegetative Any part or condition of a plant not associated with flowering. A vegetative shoot, for example, does not produce flowers; nor will a juvenile plant.
Veneer grafting *see* page 178.
Viability A measurement of those seeds that are alive at any one moment.
Vine-eye cutting *see* page 144.
Water stress A variable condition of wilting in which plant material is losing water faster than it can take it up.
Wedge grafting *see* page 174.
Whip-and-tongue grafting *see* pages 170–3.
Whip graft A very basic apical graft made with a sloping cut on the rootstock and a matching cut on the scion. It is difficult to tie satisfactorily.
Whorl A variable group (three or more) of flowers or leaves that arises in a ring from the same point on a stem.
Worked Grafted.
Wounding *see* page 24.

Tools and equipment

To propagate plants successfully, it is necessary to have a clean and tidy working area, efficient and effective tools and kit and to follow a standardized procedure. Failure in any part of the system leads to frustration and, more importantly, delays that will reduce the probability of success.

Most important of the gardener's special tools and equipment for plant propagation are a sharp knife, a pair of secateurs, a dibber, suitable compost and a selection of pots and seed trays. Not all tools or fancy bits of equipment will necessarily enhance the success of propagation, but the important ones will because they make the gardener's job easier, and if the job is easier it often succeeds more readily.

The use of suitable tools gives the plant material the very best start. To avoid tearing and crushing, for example, always use a sharp knife or razor blade and a clean sheet of glass when preparing a softwood cutting for planting. If the plant material is damaged, it will die and become a site for possible rots to infect the cutting. By the same token it is important

not to push a cutting into the compost; always first make a hole with a dibber of suitable size, and then plant the cutting in that hole. A dibber should be approximately the same diameter as the cutting to be planted.

Although many people will use a kitchen table, draining-board or greenhouse bench, the most suitable place to make cuttings, graft or sow seeds is a bench in the garden shed with a convenient shelf for all the bits and pieces of kit, tools, rooting powders, etc. The height of the bench will be a critical factor to the comfort of the gardener if considerable time is to be spent propagating or potting plants. Incorrectly sited benches will encourage or enhance backaches and cricks in the neck. It is also important to have good lighting placed directly over the work bench itself.

Plant propagation in many ways is akin to surgery, and nowhere is there more routine and standardized procedure than in an operating theatre – where all concentration is centred on the patient.

Therefore the secret of success for a gardener lies in having all the required tools and kit readily to hand and clean and in good working order, so that any technique of propagation can proceed smoothly and all concentration can be centred on the plant material.

After use it is important to clean, service and restore all kit to its correct place so that it is readily accessible on the next occasion.

Basic tools and kit

Knives (1), safety razor
   blades(2) and secateurs (3)
Sharpening stone (4)
Oil for lubrication (5)
Cleaning rags (6), solvent (7)
   and emery paper (8)
Pressers (various) for firming
   compost (9)
Dibbers (10)
Sieve ($\frac{1}{8}$ in mesh) (11)
Labels and soft lead pencil (12)
Notebook for records (13)
Polythene bags and tape (14)
Raffia, twine, etc. (15)
Split canes 12 in or 15 in (16)
Hand sprayer (17)
Watering can (18)
Fungicides (19)
Pesticides (20)
Rooting powders (21)
Panes of glass for covering
   seed trays (22)
Panes of glass for cutting (23)
Pots (24) and seed trays (25)
Composts (26) and fertilizer (27)

WORK BENCHES

To find the correct height for your work bench, stand up straight, drop your arms to the side; then raise your forearms at right angles to your body and drop your wrists – the bench should be at the height indicated by your fingertips.

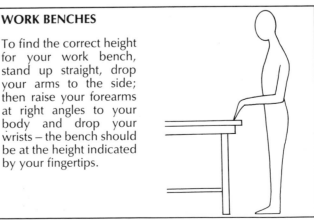

Environmental control

The main problem in propagation is to ensure survival of the propagated material (be it seed, cutting or graft) until it establishes as a new young plant. If the correct material has been used at the start, and as long as any treatments or cuts have been made correctly, then success is directly related to the control of the environment. The quicker the regenerative processes can be induced the less chance there is for things to go wrong.

In plant propagation there are two environments: the aerial environment, which can be broken down into humidity, temperature, gaseous balance and light transmission; and the environment of the medium (soil and compost), which covers temperature, moisture status, aeration and its reaction (acidity/alkalinity). Any equipment therefore should be measured in relation to the effect it has on these factors.

An ideal environment is one that allows minimum water loss from the plant material, cool air temperatures, adequate light penetration for photosynthesis, a normal atmospheric balance between compost and air, good drainage and warm soil/compost temperatures with a neutral acidity/alkalinity reaction. The degree to which a particular system of environmental control operates will limit the propagation techniques that can be used successfully within it. In general, the "softer" or less hardy the plant material the greater will be the degree of environmental control needed to achieve success. The vagaries of the normal climate are too great for all but the easiest and hardiest plants to be propagated successfully outdoors.

Cold frames

To provide initial control over the environment, place a box with a lid of glass on ordinary soil. This cold-frame environment helps to increase soil temperatures, reduce temperature fluctuation, maintain humidity and allow light penetration, and it can be used for the propagation of a wide range of hardy plants. Its main disadvantage, which is shared with all enclosed environments, is that air temperatures build up when conditions are sunny. This necessitates either airing the frame to reduce the temperature, and thereby losing humidity, or shading the glass to cut down the light input, and so reducing photosynthesis.

There are many plastic substitutes used in place of glass, but because of their heat/light transmission characteristics they are less satisfactory in the late autumn to spring period as they do not conserve heat so effectively as glass.

The most manageable cold frame to construct is made with "Dutch lights", which are single panes of glass held in separate wooden frames 4 ft 11 in long by 2 ft 6¾ in wide. These can be laid side by side across a base frame with a distance between backboard and frontboard of 4 ft 9 in. For propagation the backboard is best made at a height of about 12 in and the frontboard at 9 in. The slope of the roof should be pitched in a southerly direction. The cold frame can be made more reliable by improved sealing of any cracks in the structure and by double glazing with two layers of "lights" — the lid of a cold frame being called a "light".

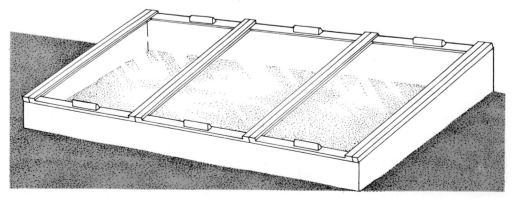

Greenhouses

The next step in the sequence of environmental control is the greenhouse, where slightly more sophisticated pieces of equipment for environmental control can be used. Greenhouses can, of course, be of a wide variety of shapes and sizes. Wooden-framed greenhouses are expensive to purchase, maintain and keep clean. Metal greenhouses are less expensive, cheaper to maintain and easier to keep clean, but unless they have an adequate internal structure they are subject to considerable distortion and damage if exposed to high-velocity winds.

A closed case, which is a frame with a lid of glass in a greenhouse, provides a high-temperature system for propagation of house plants and less hardy subjects. Accurate control of temperature can be attained by installing a thermostatically controlled soil-warming cable, which will provide bottom heat, into some sand at the base of the closed case.

Mist propagation units

The ultimate environmental control is provided by a mist unit. This is an open system that automatically maintains the moisture level while allowing the full penetration of light and the use of bottom heat without an increase in air temperature. However, such a system requires both electricity and water in the greenhouse.

Propagators

The alternative compromise is the so-called "propagator". This is a portable unit and can be used either in the greenhouse or indoors provided that adequate light is available. It consists of a glass-fibre base fitted with a thermostat and heating cables and a Perspex-type dome, which provides the closed environment. All sorts of variations are available so make sure that the propagator you buy is sufficiently large for your needs.

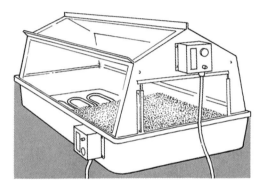

Polythene tents and tunnels

At the other extreme is a cheap and simple arrangement that provides a sufficiently effective closed environment for easily propagated plants. Place a polythene bag over the top of a pot or tray, and support it either by one or two canes or by a loop of wire with an end stuck in the compost; seal with a rubber band. Make a polythene tunnel over plants outdoors by supporting some polythene sheeting with wire and then sealing the ends.

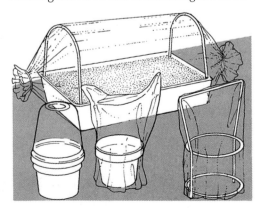

Knives

Without doubt the most important piece of equipment needed by a gardener wishing to propagate plants is a knife. Choosing it is perhaps one of the most difficult decisions to make, as so much depends on what sort of work it will be required to carry out.

For most tasks, a medium-weight knife with a sharp carbon-steel blade is best. For grafting, select a fairly heavy knife.

A budding knife has a spatula end for prizing open flaps of bark. It is a useful luxury – an ordinary propagation knife is really quite adequate. For very soft cuttings from plants such as dahlias, chrysanthemums and penstemons the best tool is a safety razor blade (that is, a one-sided blade with a thick metal-covered edge on the opposite side).

Most gardeners use a knife with a straight cutting blade, which is easy to sharpen, but some prefer a slightly curved blade. A knife with a very curved or hooked blade is not suitable for plant propagation. It is also extremely difficult to sharpen.

A knife should always be easy to open and comfortable to hold when in use. Especially with knives that may be required to cut tough material, the blade when opened should be set back into the handle. This avoids excessive play from side to side so that the blade does not loosen.

Knife blades are either hollow ground on both sides or on one side only: both types are equally efficient. One-sided knives are usually made in both right-handed and left-handed patterns. As a general rule the more expensive a knife is, the better is the quality of blade and overall design. A good-quality steel blade will maintain its sharp cutting edge considerably longer than one of poorer quality and is well worth the extra investment.

Keep a knife just for propagation and do not use it for pruning, cutting string or the

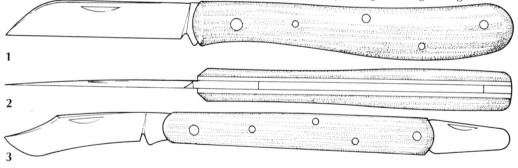

1

2

3

1 Choose a blade with a straight cutting edge as it is simple to sharpen.

2 Use a knife with its blade well set back into the handle.

3 Prize open flaps of bark with the spatula end of a budding knife.

CUTTING HARD WOOD
Hold plant material in left hand. With blade below stem and right thumb above, make a shallow-angled cut from beneath, drawing right forearm backwards and maintaining gap between right thumb and blade.

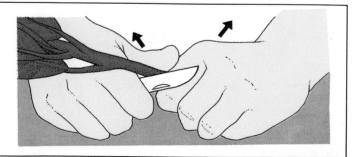

1,001 other jobs for which an ordinary pen-knife is more than suitable.

Cut soft wood against a clean pane of glass on the bench in the potting shed. Hard wood should be cut by holding the plant material in the left hand. With the knife blade below the stem and the right thumb above, make a shallow angled slice from beneath, drawing the right forearm backwards and maintaining the gap between the right thumb and the blade. Never attempt to cut by pressing the blade towards the thumb – it can have disastrous consequences for the operator.

When using a knife for cutting plant material the blade will inevitably become clogged with resins and plant juices, and these will impair the cutting efficiency. Therefore after use clean the blade either with a rag dipped in a solvent such as petrol or surgical spirit or by rubbing the blade with a fine grade of emery paper.

Sharpening a knife

To sharpen a knife successfully is often regarded as a difficult and specialist job, but in fact if a few basic rules are observed it is relatively simple provided that a straight-bladed knife is used and that the sharpening is carried out on a broad, flat carborundum stone with a coarse and fine side. The stone should be slightly lubricated with a light oil to aid easy movement.

Push the blade gently along the coarse side of the stone; then repeat the movement. Give a final "rub" on its fine side. After several turns, repeat the operation on the other side of the blade. Always traverse the entire length of the stone so that it does not become unevenly worn.

All sorts of curved and small sharpening stones are available, but for someone not used to sharpening knives they are difficult to handle and use effectively.

Sharpening a hollow-ground blade

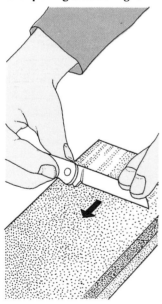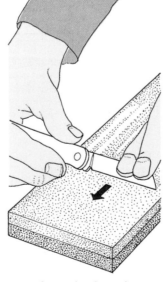

Sharpening a flat-ground blade

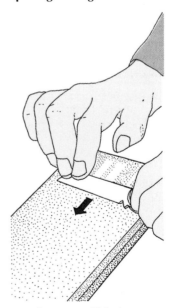

1 Pour a little oil on the coarse side of a carborundum stone. Hold a hollow-ground blade facing forward at a very acute angle to the stone.

2 Push gently along the stone. Lift, and repeat movement several times. Give final rub on the fine side of the stone.

Lay flat-ground blade on the stone. With slight pressure towards sharpened edge, draw along the stone to the end. Repeat movement.

Secateurs

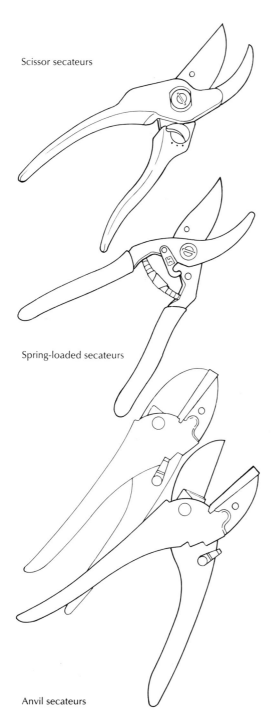

Scissor secateurs

Spring-loaded secateurs

Anvil secateurs

Although a sharp knife has always been regarded as the gardener's main cutting instrument for propagation, modern practice makes just as much use of a suitably designed pair of secateurs; and where these can be used, they are much more effective than a knife because they are quick and easy to use and because they are also less liable to cause injury to the plant material.

A pair of secateurs can not only cope with the final cut on any stem that is firm and hard enough to cut cleanly but can also be used for the initial cutting of softer stems, which may at the final stage need to be trimmed carefully with a knife or razor blade.

There are two basic secateur designs: the "anvil" type and the "scissor" type. Anvil secateurs have one sharp blade, usually hollow ground on both sides, which cuts through the stem by "crushing" it against a broad flat surface (the anvil). Scissor secateurs also have one sharpened blade, but normally only the internal surface is ground flat. This blade cuts by rotating past the anvil blade as in a conventional pair of scissors. The scissor types are preferable simply because they make a cleaner cut and cause less crushing and bruising in the region of the cut. Other secateur designs such as those with a ratchet action are not necessary for propagation.

When choosing a pair of secateurs ensure that their size is convenient, the handles feel comfortable, and that they are easy to operate. It is best to choose a spring-loaded pair so that they reopen automatically after each cut – it is tedious to have to keep opening the blades for each incision. Select a pair with a catch that keeps the secateurs closed when out of use so they are not dangerous and the cutting edge is protected. Some types of secateurs have a sap groove, which helps to prevent the blades sticking. Like all tools for propagation, a pair of secateurs should ideally be kept only for this purpose to prevent the blades becoming prematurely blunt and ineffective.

As with knives it is usually possible to judge secateurs on the basis of cost: expensive types are normally well designed, easy to dismantle and reassemble and have good-quality steel blades that retain their cutting edge for a long period.

Care and maintenance

Secateurs, more than most tools, require constant maintenance if they are to remain effective. After each session wipe the blades with a solvent such as petrol or surgical spirit and/or fine emery paper to clean off resins, plant juices and residues that otherwise may quickly impair the cutting edges. Then wipe the blades with light oil to prevent rusting, and similarly oil all the moving parts to keep them in good working order.

The cutting blade will need periodical sharpening to maintain its edge, but this will be fairly infrequent if a good-quality pair was obtained initially and the blades are kept clean. Usually, good-quality secateurs are also marginally adjustable to maintain a good cutting action.

The method of sharpening varies, but usually the secateurs are dismantled and the blade portion sharpened like a knife. In some makes the blade is disposable and is replaced rather than sharpened. Always keep the leaflet that accompanies the tool and follow the manufacturer's instructions carefully. If no instructions are supplied, take the secateurs to an expert for sharpening.

Making a clean cut

When using secateurs it is important to notice where the cut is actually being made; the anvil blade is often quite thick and it is not always possible to see the actual cut when making it. Also, ensure that any cut is made with the anvil blade away from the proposed cut surface so any bruising is not incorporated in the propagated material.

To make a clean cut, decisively squeeze the secateur blades together. Never force secateurs to make a cut as this merely bends the blades, damages the hinge and probably the plant material as well. When dealing with a thick, hard stem move the blades round a bit after each cut until they cut through it.

Correct cut	Incorrect cut	Cutting with anvil secateurs

Place the sharp cutting blade at the base of the material to be cut so it is flush with the main stem. Squeeze decisively.

Never make a cut with the anvil blade flush with the main stem as it will leave a snag.

Squeeze the sharp blade through the stem on to the anvil blade to make a clean cut without bruising and crushing.

Containers 1

Normally a pot is as deep as it is broad, but both three-quarter pots and half (or dwarf) pots – sometimes called pans – are available. At the other extreme particularly deep pots called "Long Toms" are obtainable. Select a container that holds only sufficient compost for the task in hand.

A vital consideration when choosing a pot is its capacity for drainage. It is not necessary to crock pots if a well-drained compost is used, but the base of the pot must contain adequate drainage holes; by the same token if a capillary watering system is used then there must be adequate holes for the moisture to rise up into the compost from the capillary medium, be it sand or matting. Only the pot shown below left has adequate drainage.

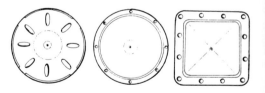

Square pots make better use of space than round ones as they can be fitted together exactly to cover an area without any waste of bench or ground space. They also generally contain a greater volume of compost relative to their surface area – a conventional 3½ in diameter round pot only contains as much

Pots

A plant pot is perhaps the commonest piece of equipment that the gardener will need for propagation. By choosing only three or four sizes in the same pot range, tasks such as watering and day-to-day management become standardized and are thus easier.

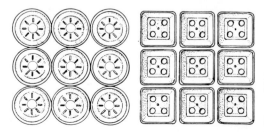

compost as a 2¾ in square pot. However, square pots are a nuisance to fill as the gardener must be sure that the compost is pushed well into the corners.

Another major initial consideration is whether to use rigid or non-rigid containers. Pots with rims tend to be more rigid than those without, and they are easier to stack. For propagation there is no substitute for rigidity, although non-rigid containers such as black polythene sleeve pots may be used at the potting-on stage. The problem with non-rigid containers is that their tendency to sag makes them a nuisance to fill.

The material from which a pot is made is also important. Traditionally, pots were always made of clay, but considerations of cost, durability and weight (in that order) have now reduced their use. Nowadays most rigid pots are made of some form of plastic and these have the advantage of being cheap, lightweight and durable. Some plastic pots, however, become brittle in time with exposure to ultraviolet light. Polypropylene pots of heavy quality will generally provide best value.

Plastic pots are also easily washed and stored, whereas clay pots require soaking, scrubbing and sterilizing between use, which is time consuming. Because clay pots are porous, the compost dries out more quickly than it would in plastic pots, and so more day-to-day management is needed.

Broad pots have greater stability, so, where a choice is available, look for a pot with a broad base and almost vertical sides. This shape also allows a greater volume of compost within the pot and therefore a more usable surface area.

Because of the problems of cost many gardeners use yoghurt cartons, vending machine cups, and cream and cheese containers instead of "proper" pots. These are quite satisfactory provided they are clean, have adequate drainage holes, and are used in conjunction with a suitable compost and management system.

DISPOSABLE POTS

It is also possible to obtain various types of disposable pots, which are usually made of some form of processed organic material. The commonest of these are compressed peat pots through which a plant's roots will pass. These have considerable value to the gardener, who can leave a plant's roots undisturbed when he transplants the whole peat pot, which will eventually rot in the ground. Pots made of paper and "whale-hide" are equally satisfactory.

The main disadvantage of disposable pots is that they are relatively expensive.

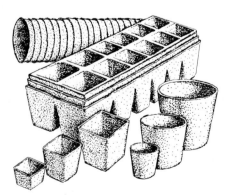

Containers 2

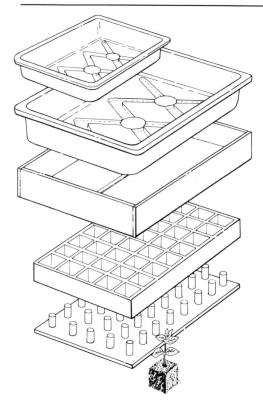

Seed trays

When dealing with relatively large numbers of seedlings or cuttings, as happens often in the production of bedding plants, a tray may be a more suitable container than a pot.

Traditionally a seed tray is made of soft wood and its dimensions are 14 in by 8½ in and, according to its required usage, either 2 in or 2½ in deep. These trays are now relatively expensive to purchase and their expected life is fairly short as they rot easily. However, they have the distinct advantage of being firm and rigid.

Plastic seed trays of the same basic dimensions are currently available in many different patterns. The most important characteristic is the degree of drainage permitted in these, and it is of paramount importance to ensure that this is adequate. The quality and variety of plastics used for making seed trays is extraordinarily variable; the best trays are those of a sufficiently thick quality to keep their shape when picked up at a corner and of a type not to become brittle on exposure to ultraviolet light. The advantage of plastics is, of course, their durability and that they can be readily cleaned. Plastic seed trays are also made in "half-tray" sizes measuring 6 in by 8½ in and 2 in deep.

Peat pellets

1 Place the pellets in a watertight container. Pour water slowly round the pellets. Leave to expand.

2 Sow the seeds or cuttings once the pellets have fully absorbed the correct amount of water.

3 Plant out as soon as roots emerge through the sides of the netting.

There are many other materials such as compressed peat, processed paper and expanded polystyrene used for seed trays, but they tend to require careful handling. Some trays are disposable, which overcomes the hygiene problem but inevitably increases the cost factor. Expanded polystyrene trays with individual growing compartments for each seed or cutting retain warmth and so promote rapid growth.

As with pots, many containers such as wooden "Dutch" tomato trays, fish boxes and moulded polystyrene packaging can be substituted for conventional seed trays provided they have adequate drainage and they are properly cleaned before use.

Peat pellets and soil blocks

As an alternative to pots and trays it is possible to substitute a system that obviates their use: the idea simply being to plant out or pot on the entire unit. This is achieved by eliminating a pot altogether, either by using a peat compost, compressed into a pellet and contained within a net, that swells up when soaked with water, or by compressing the compost into blocks.

These systems are useful, very effective and reduce the ultimate root disturbance to a minimum which means the growth of the young plants is not checked when they are transplanted.

The compressed peat pellets as typified by the "Jiffy 7" are expensive but seem to work very well, and if used for small seedlings or sturdy cuttings appear almost to enhance rooting.

The high initial cost of purchasing a soil block mould may deter many gardeners who do not propagate large quantities of cuttings and seedlings. It is best to make hexagonal soil blocks as they do not dry out so readily provided their sides touch each other. To make a soil block successfully it is important to ensure that the compost has the correct level of dampness. To test this, take a handful of moist compost. Squeeze gently but firmly; the compost should tend to crumble. If it falls apart, the material is too dry; if it does not start to crumble, it is too moist.

Fill the mould with the compost, compressing it only until the particles form a block. Place the soil block on a tray and leave it to consolidate for 24 hours before inserting a seed or cutting. As the plant grows the roots hold the block together. Plant out as soon as the roots emerge through the sides of the soil block.

Soil blocks

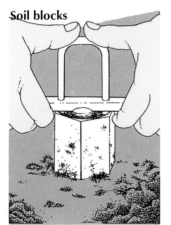

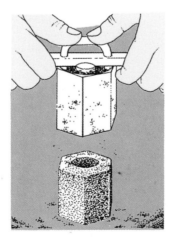

1 Hold the soil block mould with one hand on either side. Push the mould into some damp compost.

2 Release from the mould when full, and stand on a tray. Leave to consolidate for about 24 hours.

3 Insert a seed or cutting and cover with compost. Ensure that the sides of the soil blocks are touching.

21

Composts

Basically a compost is a soil substitute for propagating and establishing plants. To carry out this function a compost requires certain properties – to be well aerated, to retain water, to hold nutrients and to conduct warmth. Thus in constituting a compost the components used should be chosen to establish these particular conditions as well as maintaining them throughout the life of the compost. In order to prevent the occurrence of pests, diseases and weeds the component materials should also be sterile.

The only component of a compost that is not initially sterile is the loam. To sterilize it place the loam in a broad, flat container such as a meat tin and put in the oven at 82°C/180°F for 30 minutes. Best results are obtained if the loam is dryish and if the tin is covered with foil so that the steam generated encourages the sterilizing effect. Cool and riddle the loam before use.

Cuttings compost
The formulation of a compost for rooting cuttings really only requires two considerations: the retention of sufficient moisture to help prevent desiccation of the cutting, and the provision of an aerating agent so that air can always circulate within the medium.

Conventionally, peat has been used as the water-retentive component and, although various peats are available, sphagnum moss peat is best as it retains a good structure for a long period. For most reliable use and to achieve uniformity it should be riddled through a ¼ in sieve.

Sand is used as the aerating agent, and it also allows adequate drainage – peat by itself tending to become waterlogged. In horticultural parlance sand usually means grit, and for these purposes a washed and crushed lime-free grit providing a particle size of between ⅛ in and ⅙ in across is the most desirable. The particles should also be "sharp", that is they should not be rounded but have points and corners and thus be irregular in shape.

Although these two components provide the basic compost they can be substituted with such items as sedge peat, well-weathered sawdust, perlite, vermiculite and graded coal dust – in fact by any material that has suitable physical properties, and that is chemically inactive and biologically more or less sterile.

Cuttings composts are usually formulated by evenly mixing equal parts by volume of peat and grit, although it is often difficult to assess how much sieved peat there is in a particular mix. In the end there is no substitute to determining the "feel" of the compost and whether it has the right properties.

Compost for germinating seedlings
The composition of a compost for seedling germination does not differ greatly from that produced for cuttings, except that a little more attention needs to be paid to the nutrient and chemical aspects.

The basic components are peat and sand and for germination pure and simple this is sufficient. However if the seedlings are to remain in the compost for some time, add loam to act as a buffer in holding nutrients and controlling drying out. The amount of sterilized loam required need not be great: a formula of 2 parts by volume peat, 2 parts sand and 1 part loam is satisfactory.

As seeds are much more sensitive to the acidity in such a compost, lime in the form of ground chalk or ground limestone should be mixed in with the sand at the rate of 1½ oz per bushel of compost.

Although it is not usual to include complicated nutrient mixes in seed composts, it is important to ensure that sufficient phosphate is available. Therefore also mix ¾ oz superphosphate per bushel of compost in with the sand.

Potting composts for growing on young plants
The formulation of composts for the establishment and growing on of young plants follows on from seed composts in much the same pattern. It is necessary to prepare a compost that allows the development of a root system; contains adequate water to support the plants and sufficient nutrients not to check growth; has a suitable acidity/alkalinity status; and does not dry out too easily.

Nowadays such composts are based on the use of peat, although traditionally the John Innes composts were based on the use of sterilized loam. The recommendation of loam as a base for composts has had to be discontinued because it is no longer feasible to obtain a standard material on which a recipe can be formulated. Peat is capable of being relatively standardized and so currently forms the basis. It is important, however, to realize that loam has a steadying and controlling influence on both water and nutrient availability that peat does not provide, and so peat-based (that is, loamless) composts require a higher degree of management and maintenance. Therefore it is prudent to use loam as a minor component merely to provide the buffering action and so ease management. In practice the aim is to produce a loamless compost with added loam!

Young plants also need nutrient in the compost and this should be added at the rate of 4 oz fertilizer base per bushel of compost unless the manufacturer recommends otherwise.

There are, of course, many available proprietary brands of peat-based composts, all of which have been tried and tested successfully. Their main disadvantage is their capacity for drying out and the difficulty of re-wetting a dried compost, although this latter factor is less of a problem if a wetting agent has been incorporated. Their chief advantage is that they are ready mixed and come packed in handy-sized plastic bags.

If a peat-based compost proves difficult to rewet, then add a small quantity of wetting agent or spreader such as soft soap. Do not use washing-up liquids.

How to mix composts

The important aspect of mixing compost is to obtain an even and uniform end product. Thorough mixing of the ingredients is essential. It is also easier if you have a bushel or half-bushel box on which to base the formula as lime and fertilizers are normally added at a bushel rate. (A bushel is the amount that will fit into a box 22 in × 10 in × 10 in without compacting.)

Evenly layer the ingredients into a pile on a clean concrete floor. The lime and fertilizers should be sprinkled into each sand layer. The whole should then be well mixed with a clean shovel.

INGREDIENTS FOR VARIOUS COMPOSTS
(parts by volume)

Cuttings compost
Equal parts peat (sieved) and sand (1/8–1/6 in grade)

Compost for germinating seedlings
2 parts peat (sieved)
2 parts sand (1/8–1/6 in grade)
1 part loam (sterilized)
and 1 1/2 oz ground limestone
and 3/4 oz superphosphate
per bushel of compost

Potting composts for young plants
JOHN INNES No. 1 POTTING COMPOST
7 parts loam (sterilized)
3 parts peat (sieved)
2 parts sand (1/8–1/6 in grade)
and 3/4 oz ground limestone
and 4 oz J. I. fertilizer base
per bushel of compost

LOAMLESS POTTING COMPOST
3 parts peat (sieved)
1 part sand (1/8–1/6 in grade)
and 4 oz any fertilizer base
and 4 oz ground limestone
per bushel of compost

LOAMLESS POTTING COMPOST WITH LOAM
7 parts peat (sieved)
2 parts sand (1/8–1/6 in grade)
1 part loam (sterilized)
and 4 oz any fertilizer base
and 4 oz ground limestone
per bushel of compost

For ericaceous composts, omit lime.

SIEVING PEAT

Riddle peat through a 1/4 in sieve before mixing thoroughly with other ingredients to make the required compost.

Rooting hormones/Wounding

Certain chemicals will promote or regulate growth responses in plants when used in minute dosages, and they are used by gardeners not only for plant propagation but also to achieve a variety of other responses, such as encouraging fruit trusses to set.

These plant-growth regulating substances work at very low concentrations and within very critical limits; a substance that sets fruits at one concentration and produces roots on stem cuttings at another may be used as a weedkiller at yet another. Thus it is exceedingly important to follow dosage instructions exactly in order to obtain the desired results.

It is also important to realize that these chemicals do not constitute a panacea for success: they will not induce rooting responses if the inherent ability of the stem to produce roots is not present. Their action is merely to enhance the innate capacity of the stem to produce its roots both in greater quantities and quicker than might otherwise have been the case. If the stem cutting is propagated from a healthy plant and at the correct season, then the use of such hormones is usually of no advantage whatsoever. They should be used with knowledge, and only as and when they are likely to achieve an effect.

The majority of rooting hormones available on the market are constituted as powders, the base simply being finely ground talc. Talcum powder is used because it is extremely soft and it lacks an abrasive quality, so causing no damage to the cutting. Mixed in with the talcum powder is the rooting hormone. Normally this is a chemical, ß-indolyl-butyric acid, known as IBA. Occasionally either IAA (ß-indolyl-acetic acid) or NAA (naphthoxy-acetic acid) is substituted. The concentration for hardwood cutting propagation is normally 0.8 per cent IBA in talc; softwood concentrations are usually much less – about a quarter of this figure. All-purpose hormone powders are usually based on NAA.

In many cases fungicidal chemicals are also incorporated into the powders, so helping against any rots that may develop in the cuttings.

Rooting hormones are also made up in liquid formulations, where the chemicals are dissolved either in water or in an organic solvent such as alcohol.

It is important to emphasize that these hormones should not be used on either leaf or root cuttings. For these cuttings chemicals are not yet commercially available to aid regeneration.

How to apply rooting hormones

In order to know how to apply rooting hormones, it is important to understand one or two basic premises. Firstly, that the concentration of hormone applied to induce root formation is not the best concentration to cause root development. Secondly, although the hormone may be absorbed through the bark, most of the hormone will be taken up through the cut base of the stem cutting.

In actually applying the hormone therefore take care to touch only the basal cut surface on to the powder so that no powder adheres to the outside of the stem.

By the application of the hormone the roots are induced to form, but if they emerge and come into contact with the hormone still on the bark this may cause the roots to die off. In many cases this does not happen totally, but it may cause losses in some plants or under certain conditions; it is therefore prudent to adopt a system that is suitable for all types of plant.

If there is difficulty in getting sufficient hormone powder to adhere to the cut surface at the base of the cutting then the cutting should first be dipped in water. This is an especially valuable hint with softwood cuttings, which will benefit from the water anyway.

Make up water-based formulations by dissolving a pill in a specified amount of water. Then stand the base of the cutting in the solution for 12-24 hours. As the concentration of water-based rooting hormones is much lower than powder-based ones, the bark is not adversely affected and so the cutting can be left standing in any depth of the solution.

If an alcohol-based solution is used, dip the base of the cutting in solution and allow to drain so that the alcohol can evaporate, leaving the hormone on the cuttings.

Correct way to apply rooting hormones   **Incorrect method**

1 Dip the base of a stem cutting into water.

2 Push the base of the cutting on to the hormone powder.

3 Ensure no hormone powder adheres to the outside of the stem cutting.

WOUNDING STEM CUTTINGS

Since certain chemicals are capable of enhancing root production on a stem cutting, it is possible that other techniques may also cause a surge in natural hormone production that could improve rooting.

In some plants there exists in the stem between the bark tissues and the wood tissues a sheath of material that is capable of inhibiting root development. However, when part of this sheath is damaged, then roots will be produced normally. This damage is achieved by a technique known as wounding.

The commonest method of wounding is to remove a slice of bark from the bottom inch or so of the cutting, using a sharp knife so that the wood tissues are just exposed. Alternatively make three or four 1 in long incisions in the bark at the base of the cutting as deep as the wood tissues.

The technique of wounding can be very effective with rhododendron, daphne and juniper, but it is unwise to use it as a matter of course as it provides another possible site for infection and rotting. It may be necessary only on older, hardwood cuttings; softwood cuttings do not normally require wounding. The need to wound a cutting will be discovered only in the light of experience – a continued failure to root a cutting, which cannot be attributed to any other cause, may then suggest that the cutting may respond to wounding.

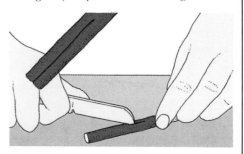

Watering

The use of water to maintain growth in potted plants is seldom efficient. The main difficulty is that the gardener is unable to water as regularly, and in the correct quantities, as the plants ideally need. A number of automatic and semi-automatic systems are available, but their operation is only as effective as the equipment and its set-up.

For propagation, water is necessary for evaporation, so increasing the level of humidity; it is also required to maintain the turgidity both of a cutting without roots and of a plant by normal uptake by the roots.

Moisten freshly potted seedlings and rooted cuttings with tepid water. Avoid cold water, which will only check growth and lower soil temperatures. If the local water is hard, it is better for the plants to use rain water.

Place a tank of water in a greenhouse where it will be warmed to the ambient temperature. The tank of water has the further advantage of providing an evaporating surface, which will help to maintain high levels of humidity in the greenhouse.

The biggest problem facing the gardener is knowing when sufficient water has been applied. To gain experience use a minimum number of pot types and sizes, and as far as possible stick to one compost formula so that experience of these limited conditions will improve watering techniques.

Watering from above with a rose, attached either to a watering can or a hose, will give a light rate of application. This helps prevent severe caking of the compost and so enhances penetration and free drainage. Water little and often, allowing the water to drain each time it reaches the rim of the pot or seed tray.

Watering by a capillary system is usually an advantage to the gardener because it is difficult to overwater, although undue drying due to poor capillary contact may prove a problem until experience is gained.

Capillary systems depend either on a proprietary matting, usually made of felt or glassfibre, or on a sand base. The former is easier to set up, but sand provides a more reliable contact. Line a tray that has raised sides with polythene sheeting. Puncture the polythene several times in a line around the sides within $1/2$ in of the required surface level. Fill the recess with fine sand to the top, and

Four methods of watering by capillary action

Choose a tray with raised sides and good drainage. Line it with polythene sheeting. Stab holes in the

sheeting about $1/2$ in below the required surface level. Fill the recess with fine sand and water to the top.

Set the containers firmly on the capillary bed so there is no air between the bed and the compost.

pour water over it. The holes in the polythene will ensure the water does not reach higher than ½ in below the surface level of the sand.

Make sure that any containers are set firmly on to the capillary bed so that the water can pass into the pot or seed tray without undue hindrance from air space between bed and compost.

If a compost is overwatered, allow it to dry out before watering again.

Should a peat-based compost dry out and prove difficult to rewet, then add a drop or two of soft soap to the water to improve water penetration. In many peat-based composts a wetting agent is incorporated and a rewetting problem should not arise.

WATERING FROM ABOVE THE COMPOST

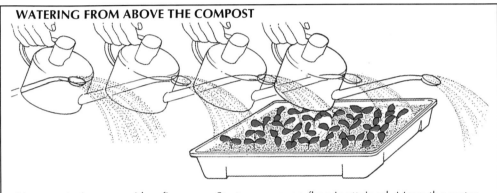

Use a watering can with a fine rose. Start pouring the water away from the container. Direct it over the compost once an even flow is attained. Move the watering can away from the container before stopping the flow of water.

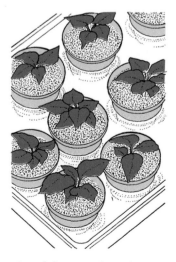

Stand the containers in a shallow bath of water. When thoroughly wet, remove and leave to drain.

Push one end of absorbent tape into a bucket of water. Lay the other end on some peat under the plant pots.

Thread absorbent tape into the bottom of a container. Place with the tape hanging into some water below.

27

Fertilizers

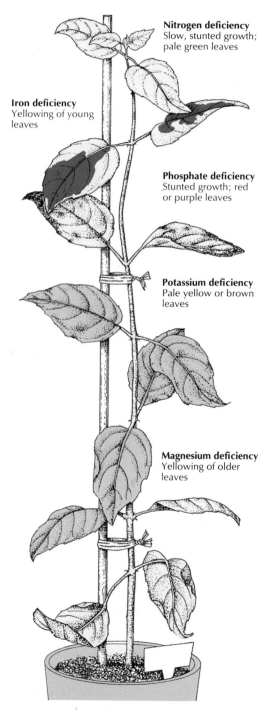

Nitrogen deficiency
Slow, stunted growth;
pale green leaves

Iron deficiency
Yellowing of young
leaves

Phosphate deficiency
Stunted growth; red
or purple leaves

Potassium deficiency
Pale yellow or brown
leaves

Magnesium deficiency
Yellowing of older
leaves

There is much misunderstanding of the role that plant nutrients play in the growth of the plant and hence they are often misused in the propagation and establishment phases of plant production.

The three so-called major elements that are required for successful plant growth are nitrogen, potassium and phosphorus. These materials are needed in addition to carbon, hydrogen and oxygen as the basic building blocks of plant material. At a secondary level elements such as calcium, sulphur and magnesium are needed in fair quantities, while trace elements, which include iron, manganese, boron, molybdenum and cobalt, are used by the plant in small to minute quantities.

Nitrogen is required wherever and whenever plant growth is anticipated. A shortage of available nitrogen is typified by the slow, stunted development of the plant and pale green leaves. It is usually taken up by the plant in the form of either nitrate or ammonium salts. Nitrogen's role in the plant is predominantly as a basic ingredient of protein and it is thus a necessary feature of developing new plant tissue. Often it is referred to as being "important for leafy growth", which is essentially true because leaves constitute a major part of plant tissue; however it is equally a necessary component of stem, root, flower and seed production. As a protein component it is also a vital feature of chromosome development.

The role of potassium in the plant is rather less easily explained. Potash, as it is commonly called, is needed as a catalyst wherever chemical reactions occur. It is especially associated with the food-making process of photosynthesis and with supplying nourishment around the plant. This again explains its general association with "leafy growth". However potash is equally important in virtually all parts of the plant where chemical reactions are occurring. Potassium deficiency in plants is usually manifest when the edges of the leaves turn pale yellow, and as this discoloration moves inwards the outer edges turn brown and appear scorched.

Phosphorus, which is normally used in the form of phosphate, has two major roles to fulfil in plant growth. Firstly it is an essential component of those very specialized proteins that constitute the chromosomes. Secondly it is the basis on which the energy needed for plant growth and development is collected, transported and released within the various chemical reactions of the plant. A phosphate shortage is much more difficult and uncertain to describe especially when it is only a marginal amount, but generally stunted growth associated with a purple or red leaf discoloration is a typical symptom; however a similar situation often arises with root damage caused by pests or rots.

Most other plant nutrients occur in sufficient quantities as minor components or impurities in the main fertilizers, and they do not specifically need to be applied as individual fertilizers.

The only two nutrients that may cause problems are magnesium and iron. The prime role of magnesium is in the formation of chlorophyll, the green colouring in the plant. A lack of it is typified by a yellowing of the older leaves as the plant transfers magnesium from its older parts to its newly created parts so causing the "chlorosis" in the old leaves.

Iron has a similar function to magnesium but it is not reusable in the plant. Its deficiency causes the young leaves to turn yellow although the veins remain green, and by this characteristic it is possible to distinguish a shortage of this element from a magnesium deficiency.

The correct use of fertilizer
It is important to ensure that sufficient nutrients are available to young plants. If composts are correctly formulated they should contain an adequate amount. However, seedlings, for example, are germinated in a compost containing only phosphate; as soon as they begin to show green leaves they will benefit from feeding with nitrogen and potash to encourage growth.

Although it is possible for the gardener to make up his own soluble feed it is far simpler and much more reliable to use one of the several proprietary brands of liquid feeds that are readily available. If a plant shows signs of, for example, potash deficiency, buy a proprietary brand of liquid fertilizer with a high balance of that particular nutrient, and not one with only that nutrient in it, and use it as recommended.

Organic fertilizers such as bone meal, dried blood and hoof and horn are too slow acting to have any real beneficial effect on a plant with a nutrient deficiency.

In the closed environment of pot or seed tray, inorganic fertilizers such as nitrate of soda, sulphate of ammonia, sulphate of iron, sulphate of potash and superphosphate should also be avoided as they may have too drastic a chemical effect on the plant and also upset the balance of the other nutrients.

If the gardener is plagued with regular magnesium deficiency substitute magnesian limestone (Dolomite limestone) for ordinary limestone in the compost.

Acid-loving plants and ericaceous ones such as heathers and rhododendrons are liable to have an iron shortage and this can be treated with a chelated iron compound either in the compost or as a foliar feed.

When applying a foliar fertilizer, always follow the manufacturer's instructions exactly. Water the nutrients over the plant's leaves using a fine rose.

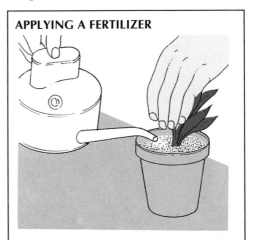

APPLYING A FERTILIZER

Make up a non-foliar fertilizer according to the manufacturer's instructions. Then water on to the compost, protecting the plant's leaves with the hand. Alternatively, place the plant pot in a bowl. Pour in a dilute solution of fertilizer. Leave overnight to absorb.

Hygiene

The greatest stumbling block to successful plant propagation is the loss of cuttings and seedlings from the action of various pests and diseases. In many cases this occurs because the gardener has failed to maintain good hygienic standards.

A system for successful propagation must be based on regular prevention and control of all possible pathogens. It is not just a question of keeping the cuttings or seedlings free of such agencies; it is also necessary to practise good standards of hygiene in the potting and propagation environment, the containers and tools used for propagation, the composts, the propagating material itself and the subsequent husbandry practised.

Always keep a scrupulously clean and tidy work bench in the potting shed. Before leaving plants to propagate in a greenhouse, scrub out all of the nooks and crannies with a solution of disinfectant so all residual infection is eliminated. It is considerably easier to do this job effectively in a modern metal-structured greenhouse than in a traditional

wooden-structured greenhouse, so take extra trouble when cleaning the latter. The best time to do this is in the early winter, when the gardener's other activities in terms of propagation are at their lowest ebb. Once clean, any remaining problems can be controlled by the routine use of various fungicide and pesticide smoke canisters, which will permeate throughout the greenhouse. At this stage, especial attention, by the use of the requisite chemical, should be paid to the control of such agents as red spider mite, whitefly, sciarid flies, mildews and damping-off fungi.

In order to avoid cross-infection in a propagating area, at the earliest opportunity always remove containers and spent compost that are not in use. Spent compost will provide a splendid home for the multiplication of both damping-off fungi and sciarid flies.

Perhaps the chief cause of infection of compost-borne rots is in the use of dirty containers for propagation. It is of paramount importance to ensure that the containers are clean, not only of fungal spores but especially of weed seeds such as chickweed, bittercress and annual meadow grass. Their source of infection usually occurs in the "crusty" layer of soil and chemicals that occurs as a tidemark on pots and seed trays. Hence the containers should be scrubbed and washed with soap solution so that they are completely clean. Clay pots will also need soaking to ensure their cleanliness. It is important to wipe all tools absolutely clean after use to ensure they do not become a potential source of infection.

The compost used for propagation must be sterile. Usually this is achieved by making up the compost from sterile ingredients. Peat, by its nature, is highly acidic and consequently to all intents and purposes is sterile. Sand should already be sterile, as will be the chemical additives. The only component that may require to be sterilized is the loam and this can be done in an oven at 82°C/180°F in a broad, flat container covered with foil so that the steam generated encourages the sterilizing process.

It is important however to remember that while all these components are sterile when fresh, if they are left lying about open to the elements they can no longer be considered to be sterile. All composts and their components should be kept bagged and covered to maintain their reliability. Incidentally, do not attempt to reuse spent compost, even if sterilized, as the chemical balances will be out of proportion.

The plant material itself must also be free of infection – do not use diseased cuttings or grafting scions for propagation. As a precaution against disease, dip leafy cuttings in a dilute solution of a systemic or copper fungicide. After they have been planted, it is a good precaution to water with another dilute solution of fungicide. Similarly, germinating seeds should be sprayed with a copper fungicide in an attempt to reduce damping-off diseases to a minimum.

As cuttings and seedlings develop, regularly use aerosol sprays or smoke canisters of fungicides and pesticides as a routine precaution against possible infections of damping-off diseases and mildews, and infestations of red spider mite, whitefly and sciarid flies.

TIDEMARKS ON POTS AND SEED TRAYS

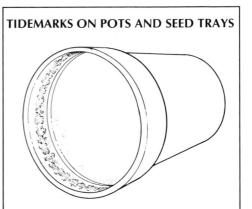

Scrub all containers thoroughly before and after use so they are completely clean and all possible sources of infection are eradicated. Remove all crusty layers of soil or chemicals on the sides of containers. Clay pots should also be soaked to ensure they are clean.

Leafy cuttings

1 Dip leafy cuttings in a dilute solution of a systemic or copper fungicide, to prevent disease.

2 Fill a pot with compost. Dibble in a cutting. Then drench the compost with a dilute solution of fungicide.

Germinating seeds

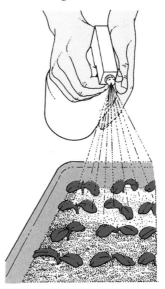

3 Spray seedlings that have recently germinated with a copper fungicide.

Pests

Various pests may cause problems when propagating, but as they attack a wide range of host plants they need only be considered in a general way so that the gardener can recognize them and treat them appropriately.

When infested propagating material, or a plant, is placed in a propagating environment, the increased levels of temperature and humidity often cause a population explosion.

It is important, therefore, to propagate, whenever possible, from material that appears to be free of pests. It is safer, and often easier, to control pests on stock plants before taking cuttings rather than to treat the cuttings later before they have established themselves as plants. This is not always possible, however, and routine pest control measures should always be taken in the propagating area to combat invasions of pests from infested plants elsewhere in the greenhouse or garden.

Aphids

Insects, such as greenfly and blackfly, are invariably present in small populations on almost all plants during the growing season, and it is important to control them not only because they can debilitate plant material very quickly with a rapid population build-up but also because they may carry virus diseases.

In a propagating area the most useful method of applying insecticides is by spraying, preferably in the evening. The most effective are those containing pirimicarb, pyrethrum, bifenthrin, permethrin or pirimiphos-methyl. Do not spray when plants are exposed to direct sunlight.

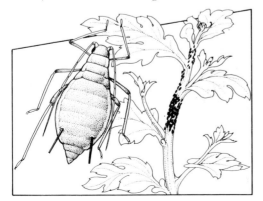

Red spider mites

These are not insects but are related to spiders. They have eight legs and are capable of spinning webs. Most are minute and comparatively difficult to see, but because they normally occur in such vast quantities their collective action is readily apparent.

A typical sign of their presence is for a yellow mottling to appear on some leaves. This discoloration gradually turns a rusty brown and is followed by a greyish sheen of web formation.

It is in the propagating environment with its increased temperatures that populations of red spider mites can build up unchecked, and they can cause problems on plants as diverse as cucumber, cyclamen, dahlia and conifers.

Available methods of control are not always effective as pesticide resistant strains of the pest may be encountered. Keep populations to a minimum by reducing the mites' hibernation stage by routinely scrubbing down all propagating areas during the winter. However, outbreaks will occur, and spraying at regular intervals with chemicals containing dimethoate, pirimiphos-methyl or bifenthrin, or the use of derris in liquid formation, will reduce infestations.

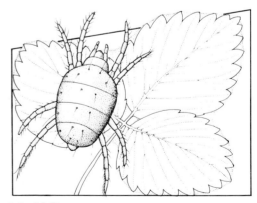

Sciarid flies

The movement of these very small black-bodied flies is more eye-catching than the flies themselves. They are probably associated with the increasing use of peat-based composts. The adults are attracted by peat, especially when it is wet, and they will lay their eggs in it. The small white grubs then proceed to eat whatever is available – in this situation usually the young, fresh succulent roots of a cutting or seedling.

Most damage is caused when the compost is overwatered, so more eggs are laid and more roots are consumed. Therefore to control infestations of sciarid flies do not allow the compost to be continually soaked. Water little and often.

The adult flies can be trapped by hanging sticky yellow strips above the potting compost. Biological control with a predatory mite, *Hypoaspis miles*, is available. Cuttings and transplanted seedlings can be protected for up to 12 months by growing them in a proprietary peat-based compost containing a slow-release formulation of imidachloprid.

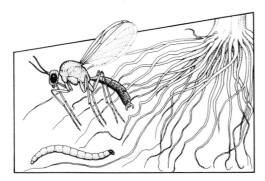

Vine weevil
The larvae of vine weevil do much more damage than is suspected, and although they are not just pests of propagation areas they can nevertheless cause havoc in newly potted young plants. If seedlings or young plants collapse suddenly during autumn to spring, they may have been attacked by vine weevil. The adult weevil lays its eggs in soil or compost. The grubs are white with a brown head, and they feed on nearby roots. Biocontrol by means of the pathogenic eelworm, *Heterorhabditis megidis*, will control vine weevil grubs if applied in late summer when the compost is sufficiently warm. Most amateur insecticides are ineffective against vine weevil. A proprietary peat-based potting compost

containing a slow-release formulation of imidachloprid will protect plants for up to 12 months.

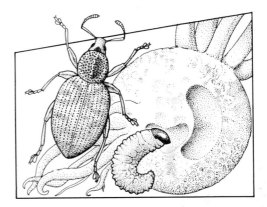

Glasshouse whitefly
This is one of the most troublesome pests of plants grown under glass. The small, white, moth-like adults and the greenish-white scale-like larvae that occur on lower leaf surfaces both feed by sucking sap, and they excrete a sugary substance known as honeydew. This makes the foliage sticky and allows the growth of a black, sooty mould. This pest is fairly tolerant of most insecticides, but pyrethrum, permethrin or pirimiphos-methyl, or insecticidal soaps may give control.

Slugs and snails
Slugs and snails feed on the seedlings of many crops. Routine use of slug bait should keep damage to a minimum. If there is a sudden infestation treat with metaldehyde or methiocarb-based pellets.

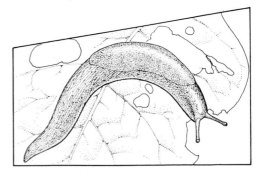

Diseases

A propagating environment often provides suitable conditions for the establishment and spread of certain diseases, but these should not become a major problem provided that correct hygienic procedures are followed. It is important for the gardener to recognize the symptoms of these diseases and to know how to apply the appropriate control measures.

Grey mould
This disease is caused by a fungus called *Botrytis cinerea*, from which it obtains its other common name, botrytis. It is an organism of general occurrence and is capable of varying degrees of parasitism. It thrives in cold, damp conditions, and it can be recognized by a brown area of rotting anywhere on a seedling or cutting, which eventually grows a covering of greyish mould. Once it obtains a foothold it is by no means easy to control its spread. To avoid an outbreak of grey mould, place any seedlings and cuttings in a well-ventilated warm atmosphere and maintain strict levels of hygiene. If an infection occurs, the affected plants should be picked out carefully and destroyed. The remaining plants should then be sprayed with carbendazim to cut down any further spread of the disease, but too regular use of this fungicide could lead to the build up of resistant strains of the fungus.

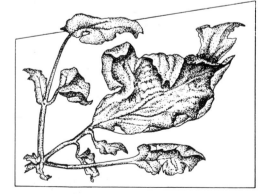

Damping-off diseases
These are usually caused by a group of closely related fungi, although certain other organisms can cause damping-off. These fungi cause problems to germinating seed-lings when threads of the fungus, which resemble fine cotton-wool, spread rapidly in the soil or over the compost surface. As it comes into contact with seedlings the parasitic fungus penetrates the tissues, which then die. The problem is particularly acute as these fungi are always present in soil or compost and are capable of survival without parasitism. Old soil or compost, therefore, provides a source of infection and should never be re-used for seed sowing or cuttings. The effects are compounded when seedling densities are high; if temperature conditions are too hot or too cold for strong seedling development; if ventilation is poor and the soil too damp; and if the seedlings or cuttings suffer frost damage.

The disease can be recognized when seedlings or cuttings suddenly die in patches.

To avoid damping-off disease sow seeds at a correct density so that the seedlings are not overcrowded; give them plenty of light, air and warmth, and do not overwater. Spray the seedling area before germination with a solution of a copper fungicide. Further control may also be achieved after the seeds have emerged by spraying again.

If damping-off diseases do become established, the seedlings or cuttings should be sprayed with a copper fungicide, which will usually kill it off. Unfortunately copper is toxic to some plants, so check the instructions carefully before applying a fungicide containing copper.

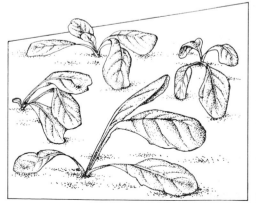

Rhizoctonia

This particular fungus may cause many problems in horticulture. It is essentially a fungus that survives in the compost and infects and destroys plant material where the humidity is too high and the conditions too warm. In plant propagation the disease normally affects cuttings inserted in mist propagators, closed cases or in pots under polythene bags. It causes basal rot of the cuttings, and it is by no means easy to control once it has established itself in the propagating area.

To prevent initial establishment of rhizoctonia, once or twice a year sterilize the propagating area. Always use equipment and containers that have been thoroughly cleaned when propagating and ensure the compost is fresh and sterile.

If the infection does appear, pick out and destroy the diseased plants.

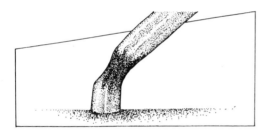

Virus diseases

Plants suffer from many diseases caused by minute living organisms called viruses. In most cases the plant merely tolerates the presence of the virus and exhibits certain symptoms indicating its presence – usually these symptoms are typified by a yellow mottling or mosaic effect on the leaves. These symptoms, however, should not be confused with the yellowing caused by magnesium or iron deficiencies (see page 29).

The effect of virus diseases is usually simply to reduce the vigour of the plant. This reduction in vigour sometimes means that a plant produces very little suitable propagating material as its growth is weaker than that of a virus-free plant of the same species or variety. It is, however, unwise to propagate from any plant that you suspect is affected by virus as the virus will be passed on to any vegetatively propagated offspring.

Unfortunately it is not possible for gardeners to reduce or eliminate the disease from a virus-infected plant. All that can be done is to attempt to prevent the spread of the disease by digging up and burning the whole of the infected plant.

Some virus diseases are carried by aphids that infect other plants when feeding; the spread of virus diseases to uninfected plants is therefore best controlled by the routine use of pesticide sprays.

Virus diseases can also be spread by eelworms in unsterilized soil, or merely by contact from, for example, a propagation knife. After cutting virus-infected plants a knife will be a fertile area for infection. Always disinfect propagating tools and your hands if virus diseases are suspected.

The only satisfactory way to avoid trouble when propagating is by using plant material that is virus free. This is a counsel of perfection as it is not easy for the gardener to recognize with certainty virus symptoms in many plants. Although viruses may be passed on through the seed it is possible to raise fresh virus-free stock of certain plants, such as *Daphne mezereum*, where viruses do not appear to be transmitted by seed. Virus-free rootstocks are labelled EMLA, and they are available for some fruit trees and a few ornamental crabs and cherries.

Some virus diseases, such as *Daphne mosaic*, can be recognized by holding the leaf up to the light and by identifying the mosaic pattern of lighter yellow colour. Others are more difficult to recognize and may frequently be confused with mineral deficiency symptoms or even pest damage.

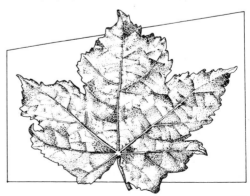

Seeds

The successful propagation of plants from seeds is a highly improbable process considering the vast number of seeds produced by the parent plant and the relatively small number of plants that survive to maturity under natural circumstances. The gardener, therefore, must recognize all the possible limitations to success, and attempt to reduce or eliminate these and so produce an acceptable crop.

Nevertheless, for the gardener, the technique of propagating plants from seed is a very worthwhile and satisfying exercise as it can be a prolific method of plant production. It is also gratifying to collect the bewildering variety of seeds in the garden without causing injury to the plants – an inevitable consequence of vegetative propagation.

A seed is produced from the fertilization of the female part of a flower by pollen from the male section. Seed is the end product of the sexual process and as such produces a population of plants that exhibit variable characteristics. By a controlled breeding programme, it is possible to eliminate the greater proportion of this variation and produce a population of seedlings that to all intents and purposes are similar. This is the usual practice in the production of bedding plants, vegetables and flower crops – that is those plants with a short enough life-cycle to allow an intensive breeding programme. Woody plant seedlings are more variable because of their longer life-cycle and their tendency to cross-pollinate in their natural habitat.

Seeds are a resting and survival stage in the continuance of a plant's existence. Basically a seed consists of an embryo, which is the young plant at its most immature and in its simplest components; a food supply, which maintains the embryo throughout the resting period and provides the basis for further development when germination gets under way; and a seedcoat, which acts as the protective component. The embryo consists of the young root system, or radicle; the young stem system, or plumule, which carries the seed leaves, or cotyledons (which may be adapted for food storage); and the hypocotyl, which is the junction between the root and shoot system. Examples of different embryos are given below: one has the food stored in the endosperm; the other has the cotyledons adapted for food storage.

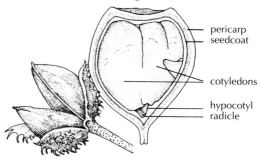

Beech *Fagus sylvatica* × 2.5

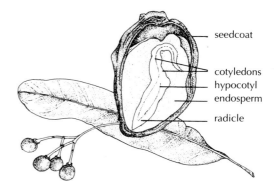

Lime *Tilia platyphyllos* × 9

There are a number of distinctions that can be made within seeds as a group.

The enormous variation in the size of seeds will inevitably influence the success with which they are propagated. Large seeds, such as acorns, chestnuts and hazelnuts, are produced in small numbers, germinate satis-

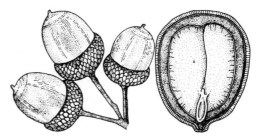

Oak *Quercus robur* × 1

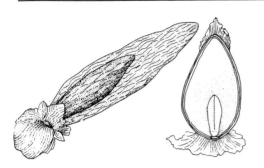

Rhododendron maximum × 24

factorily, and as a general rule establish well. Dust-like seeds, such as those from rhododendrons and lobelia, have a low germination and survival rate.

Seeds also vary greatly in the materials that they use as food reserves – that is the stored food in the seed. Those plants that store food as carbohydrates, such as elderberries, marigolds and laburnum, are generally stable and long-lived, and will withstand drying. Seeds that store food as fats or oils, for example peony, magnolia and chestnut seeds, deteriorate both with time and drying and so present problems of storage and survival. It is better to allow these seeds to mature on the plant and collect them just before dispersal.

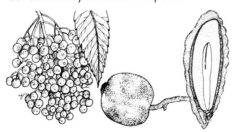

Elderberry *Sambucus nigra* × 10

Magnolia grandiflora × 2

Survival of drying, however, is not just a function of the stored food; it also reflects the condition of the seedcoat and its ability to protect the seed. The seed of plants such as willows, with very poorly developed seedcoats survives for only very short periods, while that of plants such as sweet peas, laburnum and lupin, with very hard, impermeable seedcoats usually survives for considerable periods in a wide variety of conditions. The seeds of the Indian lotus (*Nelumbo nucifera*) are reputed to have retained viability in a peat bog for over one thousand years.

The variation in characteristics of the seed, and more often the fruit, are endless; some seeds and fruits have large wings, hooks or other projections that provide an aid to dispersal, and these can easily be trimmed or rubbed off. The shape of a seed is designed so that, when it is dispersed, it will fall to the ground and lie in the best position for germination. Altering its shape may affect this characteristic, and so, when planting, try not to place a seed upside down. If incorrectly positioned the stem of the germinating seed may produce a kink.

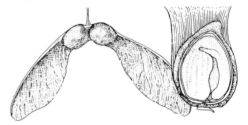

Maple *Acer platanoides* × 3

Commercial seeds

As well as being sold loose in packets, seeds are- now available commercially in other forms that make sowing easier and more accurate. Pelleted seeds are coated with decomposable material which disintegrates when in contact with moisture. It is especially convenient to buy small seeds in this form as they are much easier to handle and sow. Seeds can also be purchased evenly spaced on a tape of decomposable paper or plastic. Just cut the tape to the required length and place it in a furrow ½ in deep and then cover with soil.

Collecting and storing

Although it is usual to buy seeds it is often quite possible for the gardener to collect, extract and store his own seeds. This is especially satisfactory with seeds from trees and shrubs, which are dealt with separately on pages 54–65.

It is important to label each stock of seed at all stages. Unidentifiable seeds will be of little use, so write a non-perishable label and make sure it accompanies the seed lot through to packeting and sowing.

Annual bedding plants are selected strains that have been line-bred to come true from seed. The problem in collecting seed to come true is that a plant may have been chance-pollinated by a different variety or species, and so hybrid, atypical plants will be produced. Commercially, the seed is kept true to type by growing the parent plants in large isolated blocks. Certain plants, such as pansies, are self-pollinating and their seed can be collected with confidence that they will come true.

Collecting and drying flower seeds
Except for those seeds that are collected and sown "green", such as snowdrops and anemones, the majority of herbaceous plants are collected, dried and extracted, and stored.

Their seeds should be collected as they become ripe and before they are dispersed.

This requires careful observation. If they are enclosed in some form of fruit, the job becomes much easier, because the seed is actually completed before the fruit matures to a stage of dispersal. The only problem then is to separate the seeds from the fruit. In most cases this involves drying, either in the sun, in a dry atmosphere or in an airing cupboard.

If fruits are collected individually, they should be broken open and then spread on tissue paper in a shallow box or tray and left to dry. If whole flower heads are collected, bunch a few stems together and hang them up to dry with their heads enclosed in a brown paper bag that is lightly tied round the stems. As they dry, occasionally shake the bag so the seeds drop into it. With small flower heads leave the neck of the brown paper bag open. Place them in a warm (21°C/70°F) environment.

After drying, break up the seed capsules to free all the seed and clean the seed lot; depending on its size it can be picked over, put through a sieve, or winnowed in a breeze.

Large fleshy seeds such as those from cyclamen, lilies and hellebores will not usually respond well to drying, and it is better to allow them to mature on the plant and collect them just before dispersal.

Spread fleshy capsules on paper in a tray or box. Leave to dry in the sun until the seeds can be extracted.

Bunch flower stems together before hanging them to dry with their heads enclosed in a brown paper bag.

Tie the neck of the bag and leave it in a dry, airy place. Shake occasionally so the seeds drop into it.

Storing seeds

The longer a seed is stored the more food is used in its survival; thus less food is available for the embryo at germination, and so germination becomes progressively less vigorous. Storage conditions should keep the seed's activity to a minimum.

Seeds should be stored dry in linen bags, paper bags or packets, or cellophane envelopes; plastics such as polythene are not advisable for flower and vegetable seeds as they tend to conserve dampness if it is present. For storage of seeds from trees and shrubs see page 62.

Always keep seeds dry and store them in a cool place such as a loft, cellar or possibly a refrigerator. If there is a danger of dampness from the environment, place the packets in a polythene bag for protection.

If properly dried, most flower and vegetable seeds can be stored for two or three years at least, because they store their foods as carbohydrates. Fleshy seeds, however, store their foods as oils or fats and are, therefore, short-lived even under the best conditions: do not expect them to survive for more than twelve months. It is probably best to store these seeds at the moisture content at which they are dispersed, so place them in a polythene bag in a refrigerator.

GENETICS

Plants grown from the seed of species or stable variants will come "true", that is be similar to the parents. If, however, one parent is unstable or normally propagated vegetatively, then the offspring will in all probability be of the normal forms of species and not of the variant.

In plant breeding there is much use of the technique known as F_1 hybridization. This is an involved process in which two true-breeding species, or stable variants of the species, are crossed to produce a hybrid generation (the first filial or F_1 generation). The advantage of these hybrids is that they are often more vigorous than their parents and may have characteristics of height, form and colour that make them more desirable.

If these F_1 hybrids cross, then their offspring, the F_2 generation, will not be like the F_1 hybrids but will revert to many characteristics of the original true-breeding parents. Hence the F_2 generation will not possess all the desirable characteristics that were present in the F_1 generation and it is therefore necessary to produce F_1 seed afresh each year.

Collect small seedheads when nearly dry. Place in an open brown paper bag and leave to dry further.

Break up dried seed capsules. Clean seed lot by sieving, winnowing or picking over the detritus.

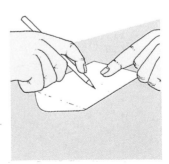

Place properly dried seed in linen bags or paper packets. Label clearly. Store in a dry cool area.

Sowing in containers

To assist with germination and the establishment of a new plant, it is often helpful to soak seeds in water for 12–24 hours before being sown in a compost that will provide adequate aeration, sufficient water-holding capacity, a neutral acidity/alkalinity reaction and sufficient phosphate. Thus a "sowing" or "seed" compost should be used.

Before choosing a pot, pan or tray decide how much seed is to be sown; the container should be large enough to allow the seedlings space to develop to the size at which they are to be pricked out.

Heap the container with compost and then, to ensure it is evenly distributed and there are no air locks, very lightly firm it to the corners and base using the fingers. Do not compact the compost.

Using a sawing action, strike off the compost with a presser board or other piece of wood so that it is level with the top of the container. Then with a presser board that fits into the container, lightly and evenly firm the compost to ¼–⅜ in below the rim, ensuring that the surface is level.

The container is now prepared for sowing. The seeds should be sown evenly over the surface either by station sowing large seeds or gently shaking small seeds direct from the packet. When shaking, keep the packet low over the compost to prevent the seeds from bouncing and giving an uneven distribution.

1 Soak large seeds in water for 12–24 hours before sowing in compost.

2 Fill a container with compost until it is heaped above the rim.

3 Firm the compost into the corners and base using the tips of the fingers.

7 Turn container through 90 degrees. Sow the remaining seeds.

8 Cover the seeds by sieving on compost, keeping the sieve low over the seeds.

9 Label the seeds with their full name and date of sowing.

If the seeds are very fine it is easier to distribute them evenly, and see where they are sown, if they are mixed thoroughly with some dry fine sand. Sow the seeds by shaking across the container, using about half the seeds; then turn the container through 90 degrees and sow the rest of the seeds in the same way.

Gently shake some compost over the container through a ⅛ in sieve so that an even and uniform layer covers the seeds. As a general rule seeds do not need to be covered by compost deeper than their own thickness.

Finally label the seeds and water them in by standing the container in a shallow bath of water so that the water moves up by capillary action. Do not stand the container in so much water that it overflows the rim on to the seeds and compost. After watering stand out to drain.

Alternatively water the compost from above, using a watering can with a fine rose. Start pouring the water away from the container and once an even flow is attained direct it over the seeds; similarly, to stop, move the water away from the container and then stop the flow, so that no drops fall on to the compost.

Cover the container with a piece of glass and place in a warm dark place, for example an airing cupboard. Otherwise, cover with glass and a sheet of paper and leave in any warm (21°C/70°F) environment.

4 Strike off the compost using a sawing action until it is level with the rim.

5 Firm the compost lightly to ¼–⅜ in below the rim using a presser board.

6 Sow half the seeds across the container, keeping hand low to prevent bouncing.

10 Water in the seeds from above the compost, using a can with a fine rose.

11 Cover the container with a pane of glass to keep the seeds moist and warm.

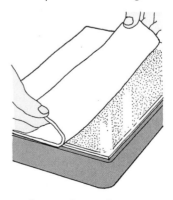

12 Place a sheet of paper over the glass to minimize temperature fluctuations.

The developing seed 1

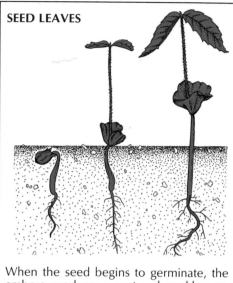

SEED LEAVES

When the seed begins to germinate, the embryo produces a root and seed leaves. These seed leaves are usually different to the true leaves that will follow.

Germination

The germination of seeds covers the entire process, from subjecting a resting seed to suitable conditions to cause it to develop to the stage at which the seedling produces true leaves and establishes as a young plant. If a seed is subjected to the conditions required for germination, and it fails to germinate, despite the fact that it is alive, then the seed is described as being dormant (see pages 55–6).

Water is vital to allow plant growth to get under way. So, if the seed has not been soaked before sowing, it is important that the compost should be watered immediately after sowing.

Once the seed has sufficiently imbibed, the embryo inside the seed begins to produce root and stem systems, which eventually break out of the seed.

To grow, the embryo uses its food reserves. When oxygen is combined with carbohydrates in these food reserves, the energy necessary for growth is produced. Thus the germinating seed will have a massive oxygen

The developing seed

1 Remove the glass and sheet of paper as soon as the seedlings appear. Place in a well-lit area.

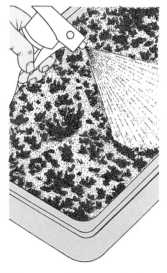

2 Spray seedlings regularly with water, but do not allow compost to become waterlogged.

3 Water in a fungicide to prevent or contain any outbreak of damping-off diseases.

requirement, which can only be satisfied by a well-aerated environment within the compost.

All growth processes within the seed are chemical reactions activated by the addition of water. To develop successfully, the seed needs an increasing quantity of water, and the compost must be capable of holding these amounts.

As all the processes involved are basically chemical reactions they will obey normal physical rules, the simplest of which implies that the higher the temperature is raised, the faster will be the rate of the reaction. In practice, this means that the warmer seeds are kept, the quicker they will germinate. As all these reactions are taking place in a biological context, there are biological limitations as to how high the temperature can be raised. In practice there are also economic considerations, because high temperatures are costly to maintain. Experience suggests that a germination temperature of 21°C/70°F is a reasonable compromise for most flower and vegetable seeds, and this is why an airing cupboard is an excellent place for seeds to germinate. For germination of tree and shrub seeds see page 68.

To keep seeds moist and warm, cover the container with a sheet of glass so that water condenses on the glass and falls back into the compost. To minimize temperature fluctuations cover the glass with a sheet of paper.

As soon as the seedlings emerge, both paper and glass should be removed. Spray the seedlings regularly with water and place them in a well-lit area, out of strong direct sunlight to avoid scorching.

Spray germinating seeds with Cheshunt compound fungicides regularly or they may succumb to damping-off diseases.

If the seedlings are to be kept in their container for some time they should be given a liquid feed diluted according to manufacturer's instructions, because many seed composts contain only a phosphate fertilizer.

Pricking out

As soon as seedlings can be handled, transplant them into a more suitable compost, leaving enough space for unrestricted

Pricking out

4 Knock the sides of the container on the work bench to loosen the compost and seedlings.

5 Loosen the compost further with the dibber, lifting a clump of seedlings.

6 Lift one seedling free of compost by holding its seed leaves and gently pulling.

The developing seed 2

development of the young plants. This is known as pricking out or potting on.

Fill a container with John Innes No. 1 compost or a compost of similar structure (see pages 22–3), and firm to the base with the tips of the fingers. Strike off compost level with the rim. Lightly firm with presser board so that the compost is ¼–⅜ in below the rim of the container, which is now prepared.

Water the seedlings; then loosen them by knocking the old container so that the compost comes away from the sides. Hold a seedling by its seed leaf and gently lift with the aid of a dibber, keeping its root system intact. Never hold the seedling by its stem.

With the dibber, make a hole in the fresh compost big enough to take the roots. Drop in the seedling and gently firm the compost back round the roots with the dibber. Repeat this operation for each seedling, spacing at 24–40 seedlings per tray.

After the tray has been filled, water in the seedlings and return them to the warm environment (21°C/70°F) so that they can re-establish as quickly as possible.

Hardening off

After the seedlings have been pricked out, they have to be gradually weaned to a stage at which they can be planted out and survive cool temperatures, fluctuating water conditions and the effects of wind without their growth rate being affected. In the plant world this process is generally referred to as hardening off.

Most seedlings will have been germinated in a protected environment during the early part of the year to produce a plant of sufficient size to be planted out as soon as the danger of frost is passed. Because so many seedlings are produced in the early part of the year, and they are not hardy, and in most gardens there is a premium on any space that provides sufficient protection, plants tend to be grown at a high density.

The problem with crowding plants together is that an increase in fungal diseases both on the stems and leaves and in the compost is likely to result; the plants tend to become spindly as they compete for light; and the varying plants have different watering needs

7 Hold the seedling in one hand. Make a hole with the dibber in the fresh compost in a new container.

8 Place the seedling in the hole and firm the compost back with the dibber.

9 Water in the seedlings once the seed tray is completed. Place in a warm (21°C/70°F) area.

so more day-to-day care and attention are needed, which is of course time consuming.

Once the pricked-out seedlings have re-established, move them to a cooler environment. For this purpose there is no real substitute for a cold frame, which should be kept firmly closed. Over the course of a few weeks increasingly air the frame during the day by raising the lid, until the frame is continually aired during the day and night: indeed the lid may be completely removed during the day if it is warm. Eventually the lid can be discarded altogether.

Frosts as severe as −4°C/25°F are sufficient to penetrate into the cold frame, so, if this level of cold is expected, provide some insulation to protect half-hardy plants. The best and most easily manageable insulation should be light yet thick; coir matting and similar materials are useful and effective.

Regularly check the seedlings in the frame to ensure that they are not drying out excessively. They should not however receive too much water. If anything it is better to err on the side of dryness rather than risk water-logging. Under these cooler conditions wet composts are increasingly susceptible to fungal root rots. Similarly, the close density of plants creates conditions under which leaf diseases are capable of taking hold.

Another aspect of seedling management is the necessity for feeding. Many pricked-out seedlings will spend several weeks in the potting compost before being finally transplanted, and there is no point in starving them and preventing them developing to an adequate size. Thus the seedlings should be regularly fed using a proprietary liquid fertilizer at the intervals stated on the manufacturer's instructions. Avoid excessive feeding as it will produce over-vigorous plants whose growth will be checked on transplanting; it will also increase the risk of disease in the cold frame.

Hardening off

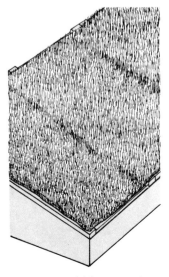

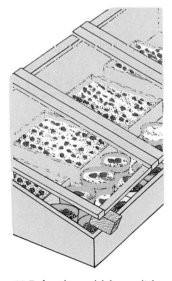

10 Cover cold frame with coir matting to insulate seedlings against damage caused by frost radiation.

11 Raise the cold frame lid to allow the seedlings to harden off.

12 Water, using a fine rose, to ensure seedlings do not dry out. Add a fungicide and liquid feed regularly.

45

Alpines

Sowing alpine seeds requires special considerations. Generally, alpine seeds are sown in autumn or winter into containers that are then stood unprotected from the elements. As the seedlings may not be big enough to handle until the following season, the kind of compost in which they are planted is important as it will have to sustain the seedlings for at least twelve months.

If the seeds have been collected, then they should, as a general rule, be sown fresh. This is especially true of such plants as *Pulsatilla vulgaris*, cyclamen, most bulbs, primulas and gentians, which show dramatically better germination sown fresh. Dried and stored seed of these plants often suffers loss of viability.

Normally, sow alpine seed in pans (dwarf pots), unless large quantities are required. Select a pan that will allow the seedlings sufficient space for growth once they have germinated.

The most important consideration will be in the choice of compost. Most alpine plants do not tolerate damp conditions, and, as the containers have to stand outside for the winter, the compost must be free draining. It therefore requires a high proportion of grit. As the seedlings may need to remain in the compost for as long as twelve months, it must retain its structure and have an ability to retain nutrients. These characteristics are found in sterilized loam, which should therefore be included. As the seedlings are generally small and fine rooted, a coarse, lumpy compost should be avoided.

The most satisfactory compost for alpine plants is made up of equal parts (by volume) sieved peat, grit and sterilized loam, with ³/₄ oz superphosphate and 4 oz lime added per bushel. Ensure the ingredients are thoroughly mixed.

Heap the pan with compost, firm very lightly to the base and then strike off. Push the compost down with a presser board, ensuring the surface is level for sowing. Mix very fine seeds such as *Ramonda myconi* with fine dry sand and sow with the packet close to the compost. To achieve an even distribution, take half the seeds and sow across the pan; turn the pan through 90 degrees and repeat the action with the remaining seeds. Large seeds such as *Cyclamen hederifolium* are station sown. With the presser board, push the seeds into the compost so they make close contact with it. Cover the seeds with handfuls of grit; then strike off level with the rim of the pan.

The seeds, while outside during the winter, will be protected by the grit, which evenly

Sowing alpine seeds

4 Sow large seeds individually by hand, spacing evenly on compost.

5 Push the seeds into the compost using the presser board.

6 Cover the seeds with grit, keeping the hand low over the pan to avoid bouncing.

filters any rain onto the compost, provides excellent surface drainage, prevents the development of mosses and algae and makes it easy to remove any weeds.

Label the pan, water and stand on any well-drained surface. While standing out, the damped seeds are exposed to the cold, which chills the seeds. This causes the break-down of those conditions causing dormancy in many alpine seeds (see pages 55–7).

Germination will occur in the spring. It is unlikely, in many cases, that the seedlings will be big enough to prick out until the autumn or following spring. Therefore give the seedlings a regular liquid feed at the intervals recommended by the manufacturer.

Preparing the compost

1 Select a container with excellent drainage. Fill with compost; then firm.

2 Strike off the excess compost until it is level with the rim.

3 Firm with a presser board until compost is 1/4–3/8 in below rim of container.

4 Strike off the grit level with the rim of the container. Label.

5 Water the seeds. Place the container outdoors on any well-drained surface.

6 Apply a regular liquid feed to seedlings once they have germinated.

Bedding plants

The sowing of bedding plant seeds is a relatively simple procedure and can be guaranteed, within reason, to produce a reliable and uniform crop of seedlings.

The main stumbling block that the gardener will experience is knowing when to sow particular plants. The object is to produce, at the same time, all bedding plants at the requisite size for planting out so that they will make a significant impact when in flower.

The sequence of bedding plant sowing is governed primarily by the speed of germination and subsequently by the growth rate of the seedlings of each species. Thus slow-developing plants are sown early in the year

Sowing bedding plant seeds

1 Mix small dust-like seeds with some dry, fine sand to extend the seeds.

2 Broadcast sow the seeds thinly, keeping hand close to compost surface.

3 Sieve just enough compost over the seeds to make an even cover.

4 Label the container with name of plant and date of sowing.

5 Water in a fungicide to prevent damping-off diseases.

6 Cover container with a pane of glass. Stand in the warm (21°C/70°F).

– although it is important to remember that in the very early part of the year light intensity is usually poor and growth rates of seedlings will be proportionally depressed.

Sow bedding plant seeds in a pan (dwarf pot) or a seed tray depending on the quantity of seed used. Because the seeds germinate relatively rapidly at warm temperatures, peat-based composts are quite satisfactory, and there are many proprietary brands available. Fill the container with compost, but keep it light and uncompacted so that drainage is maintained. The main problem with peat-based composts is that they tend to waterlog easily, causing both death of the seeds and poor seedling development – symptoms often associated with damping-off diseases and sciarid fly attacks.

Many bedding plants, for example lobelia and *Begonia semperflorens*, have incredibly small, almost dust-like seeds that are difficult to sow evenly and at a sufficiently low density. These should be thoroughly mixed with some dry, fine sand, so that an even distribution can be achieved. Covering with compost is not then necessary.

With the seed packet close to the container sprinkle the seeds evenly over the compost. Sow thinly to avoid having overcrowded seedlings later on. Sieve just sufficient compost over the seeds to make an even cover. Label the container and water the seeds either by using a fine rose on a watering can or by standing the container in a basin of water. Cover with a sheet of glass to conserve moisture and place in a warm environment (21°C/70°F) to promote germination; at this stage light is not important.

As soon as the seedlings emerge, place them in the light to encourage growth and remove the cover because excessive humidity among seedlings at this stage will lead to damping off.

The temperature, however, should still be maintained at as warm a level as possible to encourage quick growth to a size at which the seedlings can be pricked out.

As soon as the seedlings are large enough to handle, prick them out into individual pots or seed trays. Their growth will inevitably be checked when they are transplanted, but the smaller and less branched the root system is, the less damage and therefore check will be experienced.

The main problem will be the likelihood of losses through damping-off diseases. This can be prevented only by scrupulous standards of hygiene, light sowing densities and regular sprays with fungicides such as copper-based chemicals.

10 Remove glass as soon as seedlings appear. Place container in a well-lit area.

11 Prick out seedlings into individual pots once they are large enough to handle.

SEQUENCE OF SOWING

January to February
Antirrhinum;
Begonia semperflorens

February to March
Dahlia; Petunia;
Salpiglossis; Salvia

March
Ageratum; Alyssum;
Dorotheanthus;
Lobelia; Lobularia;
Nemesia; Scabiosa;
Tagetes – African

March to April
Callistephus; Nicotiana;
Portulaca;
Tagetes – French

April
Zinnia

Herbaceous plants

The growing of herbaceous plants from seed is by no means as widely practised as the growing of bedding plants or even alpines from seed. This is largely because most herbaceous plants are selected forms that require vegetative methods of propagation. However, there are many herbaceous plants, such as delphiniums and lupins, that can be grown successfully from seed.

Most hardy herbaceous plants, and especially those that disperse their seeds in the late summer and autumn, will produce seeds that require a period of exposure to the cold to break their dormancy. Their seed is sown in autumn or winter in containers that are then left outdoors; germination should occur in the spring.

Those plants that disperse their seeds in summer, after an early spring flowering, often do not show any dormancy conditions, especially if they are collected and sown slightly green. This kind of plant will then germinate quickly and establish a seedling before the onset of winter.

Some herbaceous plants, and particularly members of the legume family, such as lupins, produce seed with a hard seedcoat. This prevents the seed germinating until it decomposes sufficiently to allow the seed to take up water. To speed up germination, chip the seedcoat with a safety razor blade so that water can get in; alternatively, rub the seeds with a coarse emery paper or similar abrasive until the seedcoat is sufficiently reduced to allow water uptake.

Some herbaceous plants such as some lilies and peonies exhibit an unusual dormancy condition that delays seedling emergence. If the seeds are sown in the winter/spring period, the seeds germinate as the temperature warms up, but only a root system emerges. Exposure to a further winter's cold is necessary for the stem to develop. Therefore it is not until the second spring that the seedlings appear. For these plants failure should not be accepted until after the second spring; do not be tempted to throw them out if germination does not occur in the first season.

Sowing herbaceous seeds

Seeds should be sown in pans (dwarf pots) or seed trays, depending on the quantity of seed available. If the seeds are slow to germinate, fill the container with a loam-based compost, which will maintain its structure over a long period despite being exposed to natural weather conditions. Peat-based compost will suffice for seeds sown in spring.

Firm the compost to the corners and the base of container; then strike off the compost.

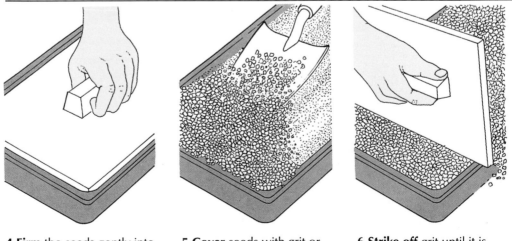

4 Firm the seeds gently into the compost with a presser board.

5 Cover seeds with grit or sieved compost, according to their requirements.

6 Strike off grit until it is level with the rim of the container.

With a presser board, firm the compost to ¼–⅜ in below the rim. If seeds are large, station sow at recommended spacing. Otherwise, broadcast sow the seeds. After sowing, firm them into contact with the compost; then cover with grit if they are to stand out for the winter, or with sieved compost if germination will occur quickly. Label the container and water in the seeds. Stand out on any well-drained surface, if the seeds need chilling.

Otherwise place container in a warm (21°C/70°F) environment.

Seedlings of herbaceous plants are susceptible to the various damping-off diseases. It is therefore important to water the emerging seedlings at regular intervals with a dilute solution of a copper fungicide.

As soon as herbaceous plant seedlings are large enough to handle, prick off into individual pots.

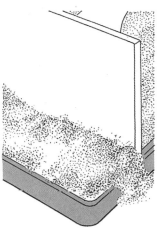

1 Heap some compost into a container. Firm gently; then strike off level with rim.

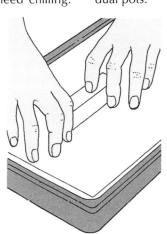

2 Firm compost to within ¼–⅜ in of rim using a presser board.

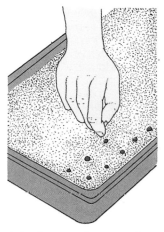

3 Station sow large seeds at recommended spacing. Broadcast sow small seeds.

7 Label container. Water in seeds. Stand container out on well-drained surface.

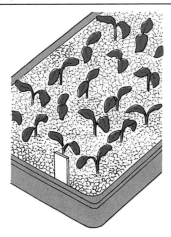

8 Apply fungicide to emerging seedlings at regular intervals.

9 Prick out seedlings into individual pots once they can be handled.

Ferns

Ferns are basically propagated from spores, and these are not the equivalent of seeds. The fern plant itself is an asexual stage in the life history of a fern: it produces spores, which are themselves asexual. The spores are dispersed and germinate, producing the sexual stage, which is called a prothallus. This is a small heart-shaped, green, scale-like growth with male and female organs, which fertilize. It is at this stage that the fern as we recognize it begins to develop.

Thus, to propagate ferns it is necessary to collect and sow spores under such conditions that the prothallus and subsequently the young fern plant will be capable of developing. Dampness and high humidity are the conditions required for the prothallus to fertilize.

As soon as a cloud of spores is produced when a fern frond is tapped, they are ripe and ready for sowing. When this stage is reached, cut the frond and place it in a large clean paper bag. Then keep it in a warm dry place for a day or so, occasionally shaking it vigorously to detach the spores, which will collect in the bottom of the bag. Never use the same paper bag twice; otherwise the spores may become mixed up with any residual ones already in the bag.

Make up a compost of 8 parts (by volume) sifted peat and 2 parts sterilized loam to give it body and 1 part crushed charcoal to keep it smelling sweet.

Choose a clean sterile pan (dwarf pot) — usually of 5½ in diameter — and lightly fill it with compost. Strike off the compost and then firm it down about ⅜ in below the rim with a presser board. Dress the surface of the compost with a thin layer of finely crushed brick before lightly sprinkling the spores over it. Then cover the pan with a pane of glass and stand it in a saucer containing soft water such as rain water. Place the pan in a warm (21°C/70°F) shaded environment and ensure that the water level is always maintained near the top of the saucer.

In three to four weeks the tiny prothalli will appear and cover the surface of the pan like a growth of liverwort or moss. At this stage the pan must be kept moist, as it is imperative that a moisture film is maintained so that fertilization can occur. Even a short period of drying may prove disastrous. As sciarid flies can also become a problem it is important to keep the pane of glass on top of the pan.

Within seven or eight weeks the small fronds of the fern proper should have appeared on the prothalli. At this stage, remove the glass to allow the fronds to develop hardily and to allow drier conditions, but still keep the pan in a warm (21°C/70°F) environment.

When big enough to handle, lift out each clump and plant it in a seed tray in a peat-based or ericaceous compost. Then grow on in a cold frame until the ferns can be separated easily and potted on.

This technique should work for most hardy ferns and indeed for many of the warm temperate ones.

1 Fill pan with compost. Add layer of finely crushed brick. Sprinkle spores lightly over the surface.

2 Cover pan with glass. Stand in a saucer filled with rain water in warm (21°C/70°F) shaded area.

3 Keep pan moist at all times. Do not remove glass when prothalli appear.

4 Remove glass once small fronds of fern develop, but still keep the pan in a warm environment.

5 Prick out clumps of ferns when big enough to handle. Plant in a seed tray and place in a cold frame.

6 Separate each fern plant when it can be handled easily. Prick out into individual pots.

Trees and shrubs 1

The propagation of trees and shrubs from seed is rewarding as it allows the gardener to practise a wide range of techniques that, if successful, produce something that has a long-lasting place in the garden or landscape. Although it is possible to purchase some tree and shrub seeds, these are limited to those kinds that can be successfully dried, so in most cases it is necessary for the gardener to collect his own seeds.

It is important to emphasize that the seeds used for propagation will only produce the kind of offspring that their heredity warrants. Seeds collected from species will probably come true; if collected from selected varieties then the seedlings will normally be of the species, unless of hybrid origin (see page 39). For this reason all fruit trees, which are highly specialized forms, should not be propagated by seed but must be increased vegetatively.

The chief problem associated with tree and shrub seeds is the presence of various kinds of dormancy, which in the extreme are sometimes combined and so present particular difficulties in getting the seeds to germinate.

The seeds illustrated on pages 54–7 are as follows. 1 *Salix* sp.; 2 *Juniperus deppeana;* 3 *Caragana arborescens;* 4 *Gleditsia* × *texana;* 5 *Cotoneaster horizontalis;* 6 *Corylus cornuta* var. *californica;* 7 *Robinia pseudoacacia;* 8 *Pyrus communis;* 9 *Koelreuteria paniculata;* 10 *Alnus sinuata;* 11 *Calocedrus decurrens;* 12 *Mahonia aquifolium;* 13 *Morus alba* var. *tatarica;* 14 *Laburnum anagyroides;* 15 *Prunus armeniaca;* 16 *Ilex aquifolium;* 17 *Fraxinus americana;* 18 *Yucca elata;* 19 *Eucalyptus fastigiata;* 20 *Populus fremontii* var. *fremontii;* 21 *Crataegus* sp.; *22 Carpinus caroliniana;* 23 *Acacia melanoxylon;* 24 *Euonymus obovatus;* 25 *Malus floribunda;* 26 *Malus baccata;* 27 *Ulmus parvifolia;* 28 *Cedrus libani;* 29 *Clematis virginiana;* 30 *Cytisus scoparius;* 31 *Viburnum lantanoides;* 32 *Taxus baccata;* 33 *Juglans cinerea;* 34 *Aesculus hippocastanum;* 35 *Rosa rubiginosa;* 36 *Ceanothus americanus;* 37 *Catalpa speciosa;* 38 *Cornus racemosa.*

Tree and shrub seeds are extraordinarily diverse in their shape and size – varying from the fine dust-like seeds of rhododendron to the large nut-like seeds of the horse chestnut or the oak, from the flat discs of wisteria to the hairy "parachutes" of clematis. All these considerations have a bearing on an individual plant's ability to survive and germinate: large seeds with a large embryo should have a much greater chance of successfully germinating than small seeds, as they have a larger food reserve, and therefore more small seeds than large seeds need to be collected.

Also affecting the quantity of seed to be collected will be the availability of seed from year to year. If this can be noted on a regular basis, it may give a guide for any storage requirements. Beech is an extreme example as it is reputed to produce good seed only once in "seven years"; other plants also have definite periodic responses.

The gardener can either purchase his tree and shrub seeds from a reliable seedsman or he can visit gardens, parks and arboreta in the hope of coming upon some unusual tree or shrub producing a crop of seeds. Gardeners are usually very generous with their plants and, if asked, will frequently be only too willing to give some seeds or cuttings.

Dormancy

If a seed is subjected to the conditions required for germination and it fails to germinate, despite the fact that it is alive, then the seed is described as being dormant.

Seed dispersed in the late summer or autumn, without an inbuilt dormancy control, would normally germinate. The seedling would then have to survive unfavourable climatic conditions that more often than not would kill it. The plant has, therefore, developed a control mechanism that prevents the seed germinating until the onset of favourable conditions for germination and subsequent establishment. Although these controls benefit the plant and enhance the chances of successful seedling production, they present a very real problem to the gardener, who either has to wait for the dormancy to be broken naturally, which can

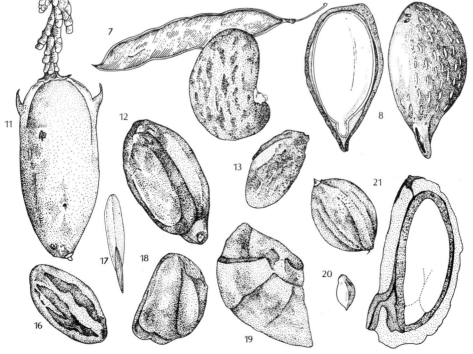

7

11

12

8

13

21

17   18

20

16

19

Trees and shrubs 2

take a long time, or has to attempt to overcome the problem artificially, which may be difficult and complex.

In woody plants there are three different kinds of dormancy.

The simplest is that caused by the seedcoat, which has thickened and hardened during the maturation of the seed. Its hardness stops water being taken up by the seed; therefore the embryo cannot imbibe and germination is prevented. In nature, this kind of dormancy is gradually reduced by bacteria and fungi in soil decomposing the seedcoat until it is no longer effective and water can be taken up.

Dormancy may also be caused by an immature embryo, which requires a warm temperature in which to develop to a stage where germination can proceed.

The commonest form of dormancy in the seeds of plants from temperate climates is a chemical inhibition to the embryo development. In nature, this dormancy is broken by normal exposure of a seed in the soil to winter's cold. This initiates a chain of events

that neutralizes the inhibitor chemical and so allows germination to proceed as soon as environmental conditions are suitable.

If only one of these kinds of dormancy occurred in a seed then overcoming it, although a problem, would be comparatively simple. Many plants, however, exhibit combinations of these dormancy controls so that overcoming them is complicated and time-consuming (see pages 62–5).

Seeds that are obtained from sources other than the gardener's own collections are almost invariably dried; the process of ripening is fully complete and all the dormancy controls are inbuilt, so germination cannot proceed until these problems have been eliminated.

However, for the gardener who is collecting his own seed it is possible to avoid the development of dormancy by collecting immature seed and preventing further drying. To do this, collect the seed when it is green to yellow to buff coloured, and fruit as it just turns yellow. At this stage the seedcoat and fruit are beginning to dry out and so develop

---

DORMANCY TABLES

Seeds with a hard seedcoat
Acacia
Caragana
Colutea
Cytisus
Gleditsia
Koelreuteria
Laburnum
Robinia
Ulex
Wisteria

Seeds requiring chilling
Alder (Alnus)
Apple and Pear
Barberry (Berberis)
Beech (Fagus)
Cherry and Plum
Clematis
Euonymus
Horse chestnut (Aesculus)
Maples – Norway and Sycamore
Oak (Quercus)
Sweet chestnut (Castanea)
Vines (Ampelopsis; Parthenocissus; Vitis)
Walnut (Juglans)

Seeds having a combination of a hard seedcoat and a chilling requirement
Cornus
Cotoneaster
Daphne
Hornbeam (Carpinus)
Magnolia
Maples – Field and Snakebarks
Roses (Rosa)
Thorn (Crataegus)
Viburnum
Yew (Taxus)

Seeds exhibiting no dormancy conditions
Catalpa
Ceanothus
Eucalyptus
Mulberry (Morus)
Poplar (Populus)
Yucca

Seeds having a hard seedcoat, an immature embryo and requiring chilling
Ash (Fraxinus excelsior)
Holly (Ilex aquifolium)

---

into the condition suitable for dispersal. It would appear that the dormancy controls develop at this stage.

By collecting the seed when it is anatomically complete with its food reserves finished but before dormancy becomes built in, the problem of the hard seedcoat is avoided and the chilling requirement is at a minimum (enough to prevent germination until the spring). Thus for germination to occur in spring, the fruits of *Daphne mezereum* should be collected not in September but in early June, when they are small, hard and green. Complicated dormancy patterns which take a long time to overcome or require a complex artificial procedure can then be avoided. However, it is easy to gather seed that is too immature, which will prove disastrous.

Once dormancy conditions are removed, the seed will germinate, provided that suitable conditions are maintained. A change in conditions, such as excessive heat or drying or a continued shortage of oxygen, will cause the development of secondary dormancy, which is extremely difficult to break down.

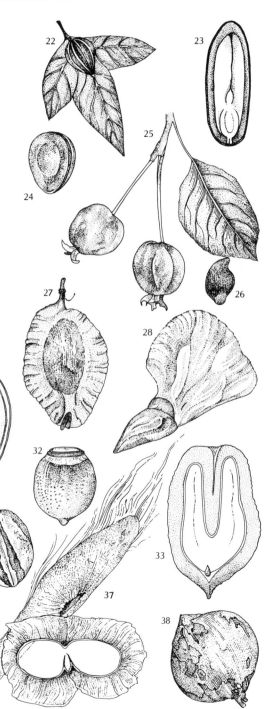

Buying and collecting

Buying seeds

When purchasing tree and shrub seeds there are few regulations which ensure the same degree of certainty in production that can be attributed to vegetable and flower seeds. Virtually no guarantees are available because the seedsmen themselves have often collected the seeds from the wild and they may be incorrectly labelled. Occasionally seed is collected which is void and so contains no embryo, even though the seed and fruit is perfectly formed when viewed externally.

Seed collectors and wholesalers usually process seeds by drying them. This has two distinct disadvantages. Firstly, the drying processes enhance the maturation of the seed and so produce deeper dormancy effects. Secondly, for those seeds that store their food reserves as oils or fats, the drying degenerates the food reserves and so produces a loss of viability. The extent to which viability is lost will depend on how much food is stored as oils or fats. Most nut seeds fall into this category.

Despite these drawbacks, there are many seeds offered by seedsmen which are entirely reliable and which are capable of surviving the drying processes without any detriment to their condition. The gardener can only learn by experience how drying affects different seeds. In many cases seedsmen do not even bother to offer seeds where there is doubt about their ability to survive the drying process.

Collecting seeds

The collection of seeds from trees and shrubs by the gardener has a number of advantages over buying in seeds.

The gardener knows the identity of the plant and that it is reasonably hardy, and he can have a clear idea of what conditions suit a particular species, having seen it growing. As so many aliens and exotics are of doubtful hardiness this is a worthwhile consideration, especially as purchased seeds probably come from a more southerly collection. A hardy parent does not necessarily produce hardy offspring, but it is more likely.

Another advantage is that the gardener can collect seeds at the moment he deems correct

Hold branch with one hand and pick with the other while collecting seeds from trees and shrubs.

Use a long-arm pruner to cut down cones. Stand at an angle so that the cones do not fall directly on to you.

— a situation that is of critical importance when green seeds are being collected to avoid a dormancy condition.

Seeds that have been collected, processed and sown without drying will not suffer losses in viability.

Finally the gardener has the advantage of choosing where he collects his seeds, and he should take them from those specimens which are regarded as desirable forms and are free from pests and diseases. Because the seeds may be the result of a chance pollination, it is not possible to expect seeds to come true, but at least, by attempting to provide a better genetic base, good forms have an increased chance of being produced.

Seeds or fruits should always be collected in prime condition, and at all stages it is important to be able to identify the seeds. Therefore label all containers used for collecting, recording not only the name of the parent plant but also the place of origin and the date of collection so that this information may be used for comparisons. The label should always accompany a seed lot until it is sown, when the information should be transferred to a permanent label. Always write labels with indelible ink.

Seeds that are green or bulky should not be collected in large batches or be kept in the containers for too long as they are very prone to "heat up" and this can very easily cause the embryo to die. These seeds and fruits should, therefore, be stored in small batches in polythene bags and kept cool in a refrigerator. Process and sow as quickly as possible.

Large cones should be cut one at a time.

When collecting seeds it is useful to have both hands free, especially if it is necessary to stretch into trees. For this purpose a collecting bag is invaluable. This can be made very simply by cutting off the top half from a plastic fertilizer or compost sack. Tie two pieces of string to the upper corners of the remaining bottom half and then secure this around the waist. This kind of collecting bag is far superior to a basket or bucket as it is not bulky and will not get in the way. Do not substitute a hessian sack because squashy and fleshy fruits may soak through the hessian into the gardener's clothing — an unpleasant and (literally) irritating experience.

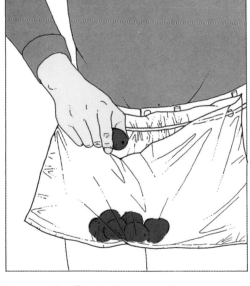

Tie a seed-collecting bag round your waist so both hands are left free to stretch into trees for possible seeds.

Label a seed lot both inside and with a tag tied on outside round the neck of the polythene bag.

Extracting

Some seeds from trees and shrubs will require to be separated from their fruit component in order to facilitate sowing and storage. In certain cases it is also necessary to store seed either from collection to sowing or from year to year. Both these factors have a bearing on a seed's viability, that is the proportion of a seed sample which is alive at any time.

The cleaning and extraction of seed from the fruit is often a tedious and time-consuming business, but it is necessary if seed is to be sown evenly.

Large dry fruits in capsules should be sieved to separate the individual seeds from the capsules. Seeds that are shed direct from the fruits, as happens with many nut seeds such as oaks, chestnuts, hazelnuts, horse chestnuts and beech, are ready for sowing. It is simply a matter of picking them up.

Winged fruits should be dried sufficiently to separate the wings, unless, as with maple, each wing has its own seed. Where the wing encloses the seed or seeds, such as with cryptic fruited plants like hornbeam, then either each seed has to be picked out or the whole fruit has to be dried and the seed separated by rubbing and winnowing.

The extraction of seeds from berries or fleshy fruits depends on the size of the individual seeds and the texture of the flesh. Pome fruits such as crab apples, pears, medlars and quinces should have the flesh pared away before any attempt to extract the seeds is made.

Separate relatively large seeds with soft flesh, for example *Berberis*, by squashing with a presser board and then swirling the material and some water around a shallow dish. The seeds will tend to gravitate to the middle and the flesh to the outside. Then pick off any berry skin that remains with the seeds.

Very fleshy fruits should also be macerated with a presser board before being left in a jar of warm water to ferment; keep in a warm place for a few days until the flesh ferments. Decant off any flesh that is floating, leaving seeds untouched in the bottom of the jar. Change the water two or three times to remove all the flesh. Then pick off any skin.

Cones of conifers are one of the most satisfying fruits to deal with. Collect the cones before they begin to shed their seeds, and place them in a paper bag in the airing cupboard. Shake regularly to separate the seeds.

Extracting seeds from very fleshy fruit

1 Pare away excess flesh from extremely fleshy fruit.

2 Place residual fruit in a sieve and squash with a presser board to break down any flesh.

3 Drop residual mass into a jar of warm water. Leave in warm place to soak for a few days.

Some cones such as those from silver fir disintegrate and then the scales have to be separated. The gardener should never open cones by placing them in the oven as excessive drying can very easily cause the seeds to die. Some cones, such as those from cedars, do not open readily in response to drying; instead put them in a saucepan containing hot water (71°–82°C/160°–180°F) and maintain this temperature until the scales open up.

Place cones in a paper bag in an airing cupboard to dry out. Shake regularly to separate the seeds.

Crumble small, dry seeds or ones that have wings by rubbing them between the hands.

Separate large seeds from their capsules by sieving out the detritus. Then pick out the seeds.

Panning fleshy fruit

4 Decant off any flesh that is floating, leaving seeds untouched in bottom of jar. Refill jar with warm water.

5 Remove the seeds from the jar once all flesh is cleaned off. Then pick off any remaining skin.

6 Squash soft flesh. Then swirl seed material and water round dish until seeds and flesh separate.

Storing/Breaking dormancy 1

STORAGE

The two main considerations concerning the storage of seeds from trees and shrubs are moisture content and temperature of storage.

As it is not always possible for the gardener, without detailed information, to know which seeds fall into which groups, all seeds should be treated similarly as a standard procedure. Surface dry all extracted seed to avoid moisture between the seeds encouraging fungal rots.

If the seeds are to be used within a couple of days, store them at room temperature in a polythene bag to maintain the moisture content at which they were extracted or collected. This is especially relevant for plants that store their food as fats and oils.

The cooler the seeds are kept the longer they remain alive and vigorous. So, for long-term storage, put the seeds in a polythene bag, label them, and then place them in a domestic refrigerator near the top, where it is coolest. The lower the temperature the more effective is storage, as long as the seeds are not frozen. Under these conditions seeds can be stored for several weeks.

1 Surface dry all extracted seeds to avoid possible fungal rots.

2 Place seeds in polythene bags to maintain their moisture content.

3 Label polythene bags with names of seeds and place in top of refrigerator.

Chipping a seed

Chip a seed with a hard seedcoat with a knife or razor blade until the seed itself is exposed and water can be taken up. Do not cut into the embryo. Alternatively, rub with a file until seedcoat is sufficiently worn away for water to be absorbed into the embryo.

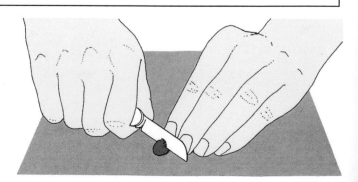

Scarification

Before sowing it is necessary to break the dormancy of seeds with hard seedcoats, and this can be done artificially by the gardener so that germination will occur as soon as conditions are suitable.

Commercially, the dormancy period is reduced by adding concentrated acid to the seedcoat, which then disintegrates. Simpler and safer methods are to be preferred, however. Seeds of a sufficient size to be handled individually can be chipped by cutting a portion of the hard seedcoat away to expose the seed itself. It is not necessary to reduce or cut away all the seedcoat but merely to allow enough to be removed so that water can enter. The seed will then swell and rupture the remainder of the hard seedcoat.

Smaller seeds and seeds with coats that do not lend themselves easily to chipping can have their seedcoats reduced by scarification. This is a fairly simple process in which the seeds are rubbed with an abrasive substance until the seedcoat is eventually worn down to a level at which water will enter. Possibly the easiest way to scarify seeds is to take a jar with a screw lid and line it with sandpaper. Place the seeds in the jar and then shake until the seedcoat is sufficiently abraded to allow water uptake.

Often a seed with a hard seedcoat also needs to be chilled to break dormancy. In this case the seed is subjected to a warm temperature to reduce the seedcoat before being stratified (see pages 64–5). As most seeds are shed in the autumn, this means that such seeds should be stored dry over winter in a warm environment, stratified in early summer and chilled during the following winter before being sown in the spring.

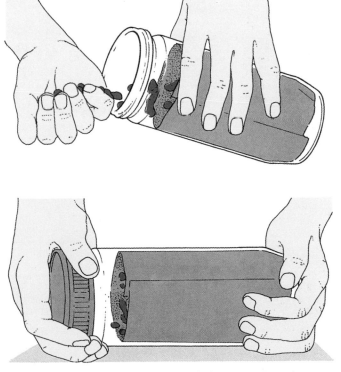

1 Take a jar with a screw-top lid and line it with a sheet of sandpaper.

2 Place a quantity of seeds with hard seedcoats in the jar and screw on the lid.

3 Shake jar until seedcoats are worn down so that water can be taken up.

Breaking dormancy 2

Stratification

One of the commonest forms of dormancy found in seeds of trees, shrubs and flowers from the temperate regions is a biochemical control of embryo development, which has to be overcome by chilling.

There are two parts to this process: the imbibition of the seed, and then its exposure to a period of cold. The simplest way to deal with the problem is to sow these seeds in the open ground, where they will receive natural chilling. In a mild winter, however, there will be insufficient cold to overcome the dormancy controls, and so germination will be delayed for twelve months following a further winter's cold.

Thus to be sure of germination an artificial treatment known as stratification is required to complete the chilling process.

To obtain a suitable medium for stratification, first sieve dry sphagnum moss peat through a ¼ in riddle. Mix about 4 volumes of this peat with 1 volume of water to produce damp peat that just exudes water when squeezed lightly in the hand. Then mix 4 volumes of this damp peat to 1 volume of seeds to give the seeds plenty of moisture. If the mixture looks compact, add 1 volume of grit to improve aeration. Place the mixture in a polythene bag and tie a label on the outside. Leave for two or three days in the warm while the seeds take up water and swell. The seeds are now ready for chilling. Place the bag in the refrigerator under the freezer box, where the temperature will be lowest (but not freezing). Turn and shake the bag every week to prevent compaction and to maintain aeration around the seeds.

The time needed for the seeds to chill is extraordinarily various and may range from three or four weeks to 16 or 18 weeks depending on the species.

Many seeds will not be bothered by an excess of chilling period as they simply sit and wait for the right conditions for germination in the spring, but some begin to germinate regardless of temperature once the required chilling period is complete.

HOT WATER TREATMENT

Some seeds with hard or impermeable seedcoats can be prepared for sowing by treating the seeds with hot water. This extracts sufficient "water-proofing" and allows the seed to take up water and swell.

Using a ratio of 3 volumes water to 1 volume seeds, place the seeds in a shallow dish and pour water that has just gone off the boil over them. Do not use more than this ratio of water otherwise the temperature will be too high for too long and this may cause damage to the embryo. Place the dish in a warm environment; leave for 24 hours. If the seeds do not swell, repeat the exercise.

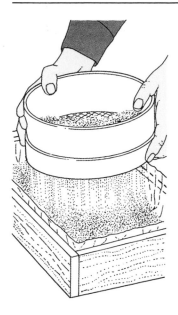

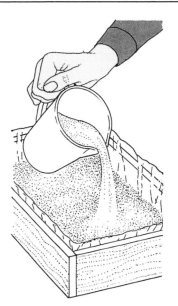

1 Sieve 4 volumes sphagnum moss peat through a riddle.

2 Add 1 volume water to the peat so it exudes water when squeezed lightly in the hand.

3 Measure out 1 volume seeds and add to damp peat mixture.

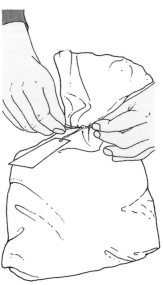

4 Mix all the ingredients thoroughly. Add 1 volume grit if the mixture looks compact.

5 Place the mixture in a polythene bag. Label and leave in a warm area for two to three days to imbibe.

6 Move polythene bag to refrigerator to chill. Turn and shake occasionally to maintain aeration.

Preparing a seedbed

An open ground seedbed is best for tree and shrub seedling propagation. Although it is quite possible to sow seeds in seed trays or pans (dwarf pots) containing compost, the quality of the compost will inevitably deteriorate and germination may be impaired after a considerable time outdoors. After germination, the seedlings should be left to establish for a growing season before transplanting, and the restrictive volume of a seed tray or pan may not allow adequate space for root or seedling development. Another drawback is that more day-to-day management is required for seedlings in seed trays and pans to ensure that watering is not sub-standard and that the seedlings have adequate nutrients.

The advantage of the open ground seedbed is that it is self-sufficient and encourages unrestricted growth of seedlings. Nor does it need to be extensive as most seedlings can be intensively grown on a relatively small area: oaks and chestnuts at ten to twelve plants per square foot; magnolias at 25 plants per square foot; and conifers at 50 to 70 plants per square foot.

The root system of many trees and shrubs is modified to live in association with a fungus that fulfils many of the functions of the root in exchange for food: this association is often obligatory for the plant and is necessary from an early stage for normal development. It is therefore important to ensure the presence of these fungi in the seedbed at germination, and leaf mould is a good source.

The preparation of the seedbed should be carried out in winter so that the soil can be left rough and allowed to weather. In order to improve drainage, soil conditions, seed covering and ease of seedling maintenance, the level of the seedbed should be raised above the surrounding soil. Set up boards 8–9 in high round the proposed seedbed. Keep the width of the seedbed relatively narrow (say 3 ft) so that seeds can be sown evenly across it. At this width, covering the seeds and general maintenance is much easier.

1 Erect side boards round proposed seedbed and fill in with soil.

2 Shovel peat and leaf mould over entire seedbed. Add grit if soil is heavy.

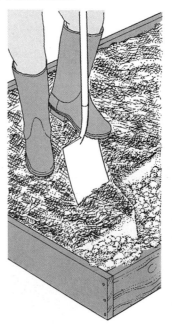

3 Dig seedbed thoroughly to a spit deep. Leave over winter to weather.

Thoroughly dig the seedbed with a spade to a spit deep, incorporating peat and, if possible, leaf mould. If the soil is particularly heavy also add grit.

In spring, knock the seedbed down to a rough tilth. This encourages weed seeds to germinate. These can then be sprayed or hoed off, so reducing the problem of weeds later on.

Prior to sowing, rake in a phosphate fertilizer at the rate of 4 oz to the square yard. Ensure the seedbed is level in order to facilitate sowing, and more especially seedbed watering. It also makes the depth of seed cover more easy to assess.

Ideal sowing conditions

The process of germination in tree and shrub seeds, as with any other seeds, should be encouraged to take place as quickly as possible so that the best use can be made of the available food reserves. Thus, tree and shrub seeds should be sown under the best possible conditions in a well-drained, aerated seedbed with sufficient supplies of water and a warm environment.

Germination is primarily dependent on water, and the seed must swell and become completely imbibed before biological activity can begin. After this stage, water is still all-important as it forms the basis of all living processes causing germination. The seedbed, however, must not become waterlogged. Air, which contains oxygen, is also essential around the seed. Energy is necessary for growth, and it is produced when oxygen is used in the breakdown of the carbohydrate reserves of the seed. Thus if conditions reduce available oxygen, germination is retarded.

The other major factor that affects germination is temperature. All growth processes are chemical reactions and as such their rate is a function of temperature: the warmer the conditions, the faster the reaction. Thus the rate of germination is directly affected by the seedbed temperature, and it is best to sow seeds when soil temperatures are warming up in the spring.

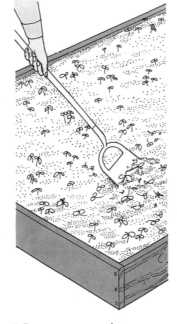

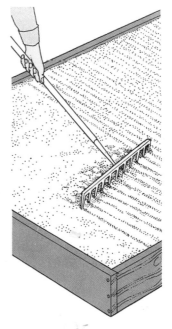

4 Break down the soil to a rough tilth in spring.

5 Encourage weeds to germinate. Then spray with weedkiller or hoe them off.

6 Rake in a phosphate fertilizer. Level soil down to a fine tilth.

Sowing in a seedbed

Sowing densities

Before sowing, it is necessary to discover the correct sowing density for the seeds to germinate and establish as healthy plants. It is as erroneous to sow too thinly and so waste space as it is to sow too thickly and have very small, useless seedlings that are liable to damp off. A desirable population will depend on the particular plant grown and its vigour.

Once the desired seedling density has been decided, this figure is modified by two factors which will then determine the number of seeds to be sown. The first is the viability of the seeds – that is the number of seeds that are alive and capable of producing new plants. Take a few seeds and, if possible, cut them open to see what percentage are viable. The second consideration that will affect the sowing density is a survival factor. As it is unlikely that all the live seeds will germinate and establish as seedlings, an estimate must be made of the likely losses, which may be caused by poor germination conditions, rots, pests and frost. In general, the larger the seed and the shorter the period between sowing and germination the greater chance it has of survival.

Sowing in a seedbed

Having arrived at a sowing rate and preferably on a still day, sow the seeds in the fine tilth of a well-prepared seedbed.

Station sow seeds that are large enough to be handled individually, spacing them evenly. To achieve an even distribution with smaller seeds, broadcast sow with your hand held just above the soil level so that bouncing is minimized. Once the seeds are sown, firm them into the seedbed so that there is intimate contact between seeds and soil and water uptake is enhanced. Then cover the seeds with grit, a shovelful at a time. Place the shovel as low over the seeds as possible,

---

SUITABLE SEEDING DENSITIES

Sow at 10–12 per square foot
Beech; Cherry; Chestnut; Nut; and Oak.

Sow at 18–22 per square foot
Crab apple; Maple; Rowan; Thorn; and Whitebeam.

Sow at 25 per square foot
Araucaria; Cotoneaster; Daphne; Dogwood; *Hamamelis; Magnolia;* Roses; *Viburnum;* and Vines.

Sow at 35 per square foot
Barberry; Holly, *Mahonia;* and *Skimmia.*

Sow at 50 per square foot
Abies; Cedrus; Picea; Pinus; and *Rhododendron.*

---

1 Broadcast sow seeds, keeping hand low to prevent bouncing the seeds.

2 Firm seeds into the soil using a presser board.

3 Cover seeds with ½ in grit, using a shovel held low over the seedbed.

and then retract it quickly, leaving the grit on the seedbed. This method reduces the possibility of the seeds bouncing. The grit, which should be about ½ in deep, will provide a well-drained surface that will allow even percolation of water through to the soil and absorb the impact of rain drops without caking and splashing: it will also keep the seeds well aerated and make it easy to remove weeds.

Finally, level the grit with the back of a rake, clearly label the seedbed and water in if conditions are dry.

This seedbed with its grit covering will maintain the seeds for an extensive period despite exposure to all sorts of weather conditions.

Protecting the seedlings
When seeds germinate they will be exposed to a number of deleterious influences which may reduce growth or even cause death. The chief of these ill-effects is wind, which causes stress in seedlings, reducing growth considerably. Watering will not necessarily alleviate the situation as the seedlings may not be able to take up water quickly enough to compensate for water lost. Therefore, shelter the seeds from wind by putting some 50 per cent permeable mesh round the seedbed.

As soon as the seedlings emerge and produce green leaves, they will need feeding with nitrogen and potash to supplement the phosphate already in the seedbed. Although these nutrients can be supplied by top dressing with a granular fertilizer, it is better regularly to use a proprietary brand of liquid fertilizer at the recommended rate, applying a small amount frequently.

Tree and shrub seedlings are extremely susceptible to damage from frost. In their natural habitat they would be protected by the woodland or scrub canopy, but in the seedbed they are completely exposed. Until the danger of frost damage is passed, therefore, they must be protected either by stretching netting with very small holes above the seedlings on a semi-permanent basis or by spreading newspaper on top of the seedlings on those nights when frost is anticipated.

In addition, it is vital to control pests and diseases such as greenfly, damping-off fungi and various mildews by spraying regularly with fungicides and pesticides.

If the seedbed has been prepared adequately then few weeds should be present in the bed. Wind-blown weed seeds, however, will appear and germinate in the grit. Pull these out while they are still small and their roots are in the grit rather than the soil.

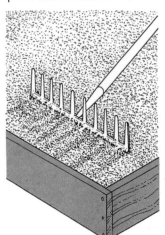

4 Level the grit by striking over with back of rake. Label seedbed clearly.

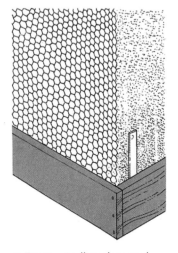

5 Erect windbreak round seedbed to reduce speed of prevailing wind.

6 Cover seedlings with newspaper when spring frosts are expected.

Exotic trees and shrubs

The trees and shrubs treated on this page are grown more for interest than for the fruit, if any, that they may produce.

Their methods of propagation are similar, except that of avocado. Fill a pot with seed compost and with the finger-tips firm gently to the corners and the base of the pot. With a presser board strike off the compost level with the rim, then press the compost to $1/4$–$3/8$ in below the rim. Sow the seeds and push them into the compost with a presser board. Cover the seed with its own depth of sieved compost. Water the pot thoroughly, label, cover with glass and a piece of paper, and place in a warm (21°C/70°F) environment.

As soon as the seedlings appear remove the paper and glass and stand the pot in good light, still in the warm. To prevent disease, spray regularly with a copper fungicide.

Prick out the seedlings once they are large enough to be handled. Knock the sides of the pot to loosen the compost. Separate the seedlings with a dibber so root disturbance is reduced to a minimum. Fill some pots with John Innes No. 1 compost or similar (see page 23). Make a hole with a dibber and plant the seedlings into individual pots; firm in, water and label. Keep the seedlings at a warm (21°C/70°F) temperature until re-established.

Citrus plants

The propagation of citrus trees and shrubs from seed is relatively simple, although it must be remembered that the resultant off spring will be of only the general type of fruit. It is necessary to propagate vegetatively if the particular variety is required.

In temperate climates most citrus fruits grow on medium-sized trees. They will not mature to that size and produce fruit in a cooler climate, where it is, however, quite possible to grow bushes that will produce fruit of a sort, provided they are kept warm in winter. The main problem is that citrus plants are evergreen and therefore need to be kept growing all the year round.

The following citrus are species and will come true: lime (*Citrus aurantifolia*); Seville orange (*C. aurantium*); lemon (*C. limon*); grapefruit (*C. × paradisi*); and sweet orange (*C. sinensis*).

Extract the pips from the fruits. Although they will withstand a limited degree of drying, it is better to sow them fresh.

Space about five seeds round a pot filled with compost. Generally, citrus seeds will germinate best at a temperature of 21°C/ 70°F and, if kept damp and warm, will appear in three to four weeks.

Coffee trees

Coffee trees can readily be propagated from seed but, like all fruit trees, the most desirable clones have to be propagated vegetatively. Seedling coffee trees do not produce reliable crops, but nevertheless most trees grown from seed do produce some fruit even when quite small: a 2–3 ft tree grown in an 8–9 in pot will flower and fruit, given suitable conditions.

A good coffee tree to grow from seed is *Coffea arabica*, which produces bright red berries that mature to a deep crimson, and contain one or two white bean-like seeds.

The seeds germinate readily if sown fresh and without drying; dried seed can be reconstituted to some extent by soaking.

Station sow about five seeds to a $3^{1}/2$ in pot; cover with compost and water in a systematic or copper fungicide. Keep at a reasonably high temperature (18°–21°C/65°–70°F).

The seedlings are particularly prone to damping off, so spray regularly with fungicide once the seeds have germinated. This should take place in about four to five weeks.

Always keep coffee trees in a warm (21°C/70°F) place and feed them occasionally.

Date palms

Most stones that can be extracted from imported boxes of dates are viable and so can be used to propagate new plants. Date palms do not fruit until they reach 15–20 ft.

Sow seeds individually in pots. Although date palms will germinate at relatively cool temperatures, they will respond more effectively and rapidly at temperatures of at least 15°–21°C/60°–70°F, when seedlings should emerge within seven to eight weeks.

Date palm seedlings are prone to damping off, so spray regularly with a fungicide.

Avocados

In common with other fruiting trees, the most reliable avocado clones require vegetative propagation. Avocado trees reach 15–20ft before they mature and produce fruit, but they are unlikely to fruit in a cool climate.

Avocado stones are large and are germinated individually in pots filled with seed compost. In a warm, humid environment (21°C/70°F) lay the stone across the top of the compost to discover which is the right and which is the wrong way up. When in about three to four weeks, the seedling's stem and roots appear, set in the right position in the soil and return to a warm, humid atmosphere.

Avocado stones left to germinate in water are difficut to transplant satisfactorily into soil so avoid this method of propagation.

Grapefruit pip   Coffee bean   Date stone   Avocado stone

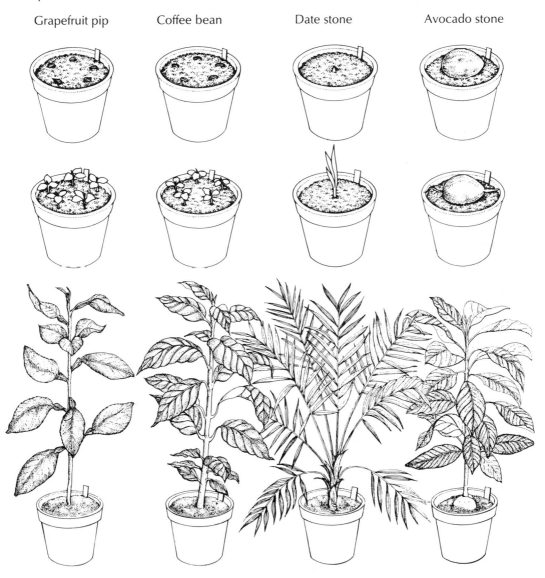

Roots

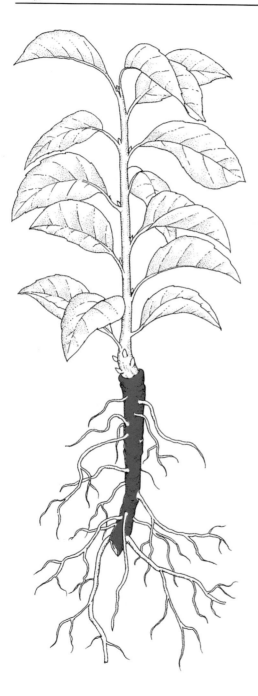

The propagation of plants from roots is a simple and rapid process that has on the whole been neglected.

As long ago as 1664, John Evelyn wrote in his famous book *Silva* of the possibility, when a tree was dug up, of leaving some of its roots *in situ* to develop as new trees. By 1731 Philip Miller, in *The Gardener's Dictionary* was describing the propagation of certain trees from root cuttings as an established practice amongst gardeners.

Since then, although it has been demonstrated that is possible to propagate in this way, the technique has never become standard, except in the case of a small number of mainly herbaceous plants. It appears to have been ignored largely because of the aura of uncertainty associated with the success of this technique. Such a technique, however, should be popular because large quantities of plants can be produced from a very small amount of propagating material. Therefore, it is necessary to sort out the relevant features involved in this kind of regeneration and determine a system that eliminates the greater proportion of the uncertainty.

Initially, it is important to divide plants into two categories: those that will propagate from roots; and those that (apparently) will not. A plant that can produce adventitious shoot buds on its roots should be suitable for propagation from root cuttings, although ultimately this is not necessarily an indication of its ability to regenerate a completely new plant.

When the various plants capable of producing adventitious buds on their roots are categorized, it will be seen that the responses vary: some plants produce buds on their roots as a natural growth process, whereas others require some agency to stimulate bud initiation. Some of them have buds that elongate and develop as shoots; others do not grow in this way.

There are three relevant methods of propagating from roots: natural suckering and division; suckers from undisturbed, but isolated roots; and root cuttings.

When a plant is lifted, inevitably some roots are severed and remain undisturbed in the soil. During the subsequent late winter and spring, these roots will, if the plant is capable, develop suckers. These suckers, if left to establish, can then be lifted and replanted at the end of the growing season. Plants that are sometimes propagated this way include *Rhus* (the sumachs), *Robinia*, *Ailanthus*, *Rubus* and *Chaenomeles*. However, most plants that regenerate in this way are also fairly easily propagated from root cuttings and, as the latter method makes better use of space, the former technique is not widely employed.

NATURAL SUCKERING

Natural suckering occurs in some plants such as lilacs and cherries. The plants send up isolated shoots that develop their own root system.

Towards the end of the growing season, sever the roots of a sucker from the parent plant and leave it to establish. After a few weeks, lift the sucker and transplant.

If a grafted plant such as a rose sends up a sucker, it should be completely removed immediately. If this is not done, the grafted plant will be weakened as the sucker is from the rootstock and not from the cultivated variety.

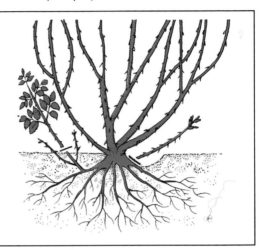

Suckers from undisturbed roots

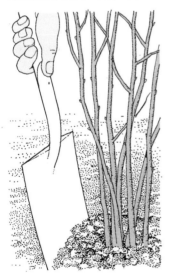

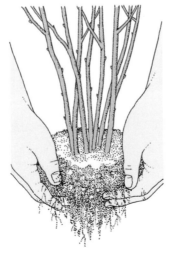

1 Dig deeply round a suitable plant in autumn. Ensure all the roots are cut right through.

2 Lift just the plant, leaving the severed roots undisturbed in the ground to establish as new plants.

3 Transplant these suckers at the end of the growing season.

Root Cuttings 1

Some plants propagated successfully from root cuttings

ALPINE PLANTS
Anchusa
Anemone
Arnebia
Carduncellus
Erodium
Geranium
Lactuca

Morisia
Primula denticulata
Pulsatilla
Verbascum

HERBACEOUS PLANTS
Acanthus
Anchusa

When to take root cuttings

When propagating plants from root cuttings it is vital to understand how seasonality affects the capacity of root cuttings to produce stem buds.

The subject of "on" and "off" seasons appears to have been virtually unconsidered until the mid-twentieth century. However, recent research has established that fluctuations in the ability of roots to produce stem buds do exist and that it is pointless to propagate while the plant's response is inhibited by adverse seasonal influences.

It is this fluctuation in the capacity to propagate that has probably produced the uncertainty which has led to the propagation of plants from root cuttings being virtually ignored by nearly all gardeners.

It is therefore necessary to determine whether the plant from which root cuttings are required does have different seasonal responses, and, if so, what is the best time to take cuttings.

Without prior guidance, the natural inclination for the gardener would be to take such cuttings in the growing season, but

experience has shown that this has met with little success; although results sometimes improved if cuttings were taken very early or very late in the season.

A few plants can produce new plants equally well at any time of the year, but these are relatively uncommon. Perhaps the best example is the horseradish, which can make itself into a pernicious weed by virtue of this characteristic: when the roots are broken, it is capable of establishing itself as a new plant from each root piece.

Virtually all other plants demonstrate a seasonal response. Early observations suggested that plants propagated most successfully during the winter, but experience has shown that, although this is typically true for woody plants, the real feature is not the winter but the dormant season. Many herbaceous plants and more especially alpine plants are not necessarily dormant during the winter. Some alpines, for example the Pasque flower (*Pulsatilla vulgaris*), start growing in the new year and if root cuttings are made after this period they do not respond; success is achieved only during their dormant

Preparing the plant

1 Lift a healthy plant from the ground during the dormant season.

2 Cut back any top growth. Shake any excess earth off the roots.

3 Wash the roots in a bucket of water or hose them clean.

Eryngium	*Romneya*	*Rhus* (sumach)	*Robinia*
Limonium		*Rubus*	
Papaver	**SHRUBS**		**CLIMBERS**
Phlox	*Aesculus parviflora*	**TREES**	*Bignonia*
Verbascum	*Aralia*	*Acacia* (mimosa) – some species	*Campsis*
	Chaenomeles	*Ailanthus* (tree of heaven)	*Eccremocarpus*
SUB-SHRUBS	*Clerodendrum*	*Catalpa* (Indian bean tree)	*Passiflora*
Dendromecon	*Daphne genkwa*	*Koelreuteria*	
	Prunus (plum and blackthorn)	*Paulownia*	

season, which is in the late summer and early autumn.

Although propagation from root cuttings is likely to succeed throughout a plant's dormant season, it is best to keep to the mid-part of that season, as a maximum response may not exist while the dormancy is still developing or once it is phasing out.

Preparing the plant
Before propagating from root cuttings, it is preferable to prepare the parent plant itself so that it will develop roots that will have a high capacity to regenerate stem buds and so produce new plants.

This ability to produce adventitious stem buds on a severed root is already present in most plants, but it can be enhanced. Lift a healthy plant prior to the growing season and shorten any top growth. Reduce its root system by cutting off the roots close to the crown of the plant, using a knife. Then return the plant to the ground. The pruning upsets the natural root/shoot balance of the plant, and it will grow quickly during the following season to bring the plant back to its normal

equilibrium between root and shoot. As a result of this treatment the vigorous, quickly grown roots will exhibit a very high level of ability to develop stem buds.

The roots will have developed fastest at the beginning of the season, and the rate will gradually have declined as the season progressed until, as the dormant season approached, growth will have ceased altogether. At the point where any root started its growth in the spring, and where it grew fastest, will also be where it has the greatest capacity to produce buds; in other words, there is a direct correlation between the rate at which a root grows and its ability to produce stem buds. If, therefore, plenty of root material is available, it is best to take a root cutting only from the top of the root where it began its development in the early spring.

In order to obtain cutting material, lift the parent plant and shorten any top growth. Wash it free of soil either in a bucket of water, or by hosing it down. It is then possible to distinguish the young roots which are suitable for propagation. Cut these close to the crown

4 Cut off the roots close to the crown, using a sharp knife.

5 Return the plant to its position in the garden.

6 Leave the plant to reestablish during the growing season.

Root cuttings 2

of the plant, at right angles to the root. Return the parent plant to the garden.

Discard the thin root end by slicing with a sloping cut. Slice off any fibrous lateral roots on the cutting to ease handling and planting later on.

The removal of roots from the parent plant for propagation will have had the additional effect of root pruning and so will cause the development of further roots for propagation in the ensuing propagating season.

Size of a root cutting

The size of a root cutting may not be critical if only one cutting is made from each root, but, with roots that regenerate readily and from which it is possible to make more than one cutting, the optimum size of a root cutting becomes very important. It is therefore necessary to determine the minimum size for a root cutting so that maximum use can be made of the available root.

The size of the root cutting depends on two factors. Firstly, the cutting requires sufficient food to initiate and develop a stem bud to the stage at which it produces green leaves and can begin to support itself. Secondly, the cutting requires sufficient food reserves to support itself while this regenerative process is going on.

The size of the cutting, therefore, is made up of the regenerative portion and the survival portion. The size of the survival portion depends on how long the cutting will take to regenerate, and this is reliant on the temperature in which the cutting is propagated: the warmer the environment the quicker the stem will develop. A root cutting taken and planted in the open ground during the winter may not produce a shoot until May, but if it had been placed in a propagator in a temperature of 18°–24°C/65°–75°F, it might well have regenerated in about four weeks. The amount of food reserve required for survival between these two temperature environments is dramatically different.

The size of the regenerative portion needed, however, will remain constant whatever the temperature of the propagating environment, so a rule of thumb measurement for the size of a root cutting is based on the variable factor – temperature.

Obtaining cutting material

 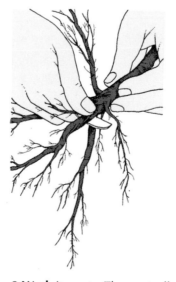 

1 Lift the plant in the middle of the dormant season. Cut any top growth.

2 Wash its roots. Then cut off the young roots close to the crown and set aside.

3 Return the plant to its usual position in the garden.

As a plant's roots will have been pruned a year before the cuttings are taken, all roots will show one year's growth and therefore be approximately the same thickness. Thus the cutting length is unaffected.

An open-ground cutting should be at least 4 in long as it will need to survive for some 16 weeks. A cold frame/cold greenhouse provides a warmer environment, and regeneration will occur in about eight weeks an so a smaller survival portion is required and the cuttings need only be just over 2 in long. In a warm 18°–24°C/65°–75°F greenhouse or propagator, regeneration time is reduced to four weeks, halving the survival time and the required food reserve once again so that in this environment root cuttings need only be about 1 in long.

Recognizing the top of a root cutting

When propagating plants from root cuttings it is very important to notice the "polarity" of the cutting – that it has a top and a bottom and therefore a "right way up". Most people suggest that root cuttings should be planted horizontally because the cuttings have been made in such a way that the top and bottom cannot be recognized, and there is no other way in which polarity can be recognized because roots have no leaves and axillary buds. However, stem cuttings are not planted on their side, so it is unreasonable to expect root cuttings to be, no matter in what direction their roots subsequently grow. Cuttings planted vertically and the correct way up will usually develop to a maximum level provided that the cuttings were taken from a healthy plant and they are given suitable conditions (see pages 78–9). Cuttings planted on their side rarely achieve more than a 40 per cent success rate.

In order to recognize the top of a root (that is, the end nearest the crown of the plant) so it can be planted the right way up, make a flat cut at right angles to the root where it was severed from its parent; at the bottom end cut away the thin portion using a sloping cut. Always cut roots in this way so, whatever subsequently happens to the root cutting, it will be possible to recognize its correct polarity and so ensure that it is planted the right way up.

Taking a root cutting

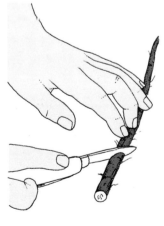

1 Cut off any fibrous lateral roots on undamaged young roots.

2 Make a right-angled cut on a root where it was severed from its parent.

3 Cut away the thin root end at the appropriate length, using a sloping cut.

Root cuttings 3

Treatment

After all the initial aspects have been considered and the root cutting has been made, the next step is to consider the question of what treatment the cutting may need in order to enhance its chances of producing a stem bud and then surviving until the bud develops and establishes as a new plant. At present there are no growth-promoting substances available for root cuttings, so that it is not possible to enhance bud production in this way. The special powders produced for inducing roots on stems should not be used on root cuttings as they will actually depress bud action.

Planting

Root cuttings need to be planted in a medium that will support them, prevent them from drying out, allow adequate aeration and, when regeneration starts, provide basic nutrients. All these features can be found in the ground outdoors and the root cuttings will do well there provided that the soil is reasonably light or they are placed under a cold frame in soil to which peat and grit have been added. However, except perhaps for a few very vigorous herbaceous perennials, it is more convenient to plant root cuttings in a container and then plant them out just as soon as they are established.

Select a container of suitable size for the number of root cuttings to be propagated, allowing 1–1½ in for each cutting. For example, plant seven cuttings in a 3½ in pot. Fill the container with a peat-based compost containing loam, which will act as a buffer to prevent excessive drying and will maintain an even level of nutrients. Strike off the compost with a presser board so that it is level with the rim. Then press the soil down to at least ³⁄₈ in below the rim of the container.

Make a hole in the compost with a dibber and then plant the root cutting. Place the top of the cutting just level with the top of the compost. Firm back the compost around the cutting. Space the remaining root cuttings evenly round the container.

Cover the cuttings with grit. Strike off with a presser board until the grit is level with the rim of the container. This weight of grit tends to compress the compost slightly so causing the tops of the cuttings to be pushed further up into the grit. This will provide almost

Planting root cuttings

1 Fill a pot with compost. Make a hole with a dibber. Plant cutting vertically.

2 Plant remaining root cuttings 1–1½ in apart. Cover the pot with grit.

perfect aeration for the bud that will develop at the top of the root cutting. Do not water. Label the container and stand it in an environment (propagator, cold frame, etc.) that is appropriate to the size of the root cutting (see pages 76–7).

Some plants, for example *Romneya coulteri*, do not like being dug up and having their roots disturbed. Therefore, place only one or two of their root cuttings in a small pot and treat as one plant, disturbing their roots as little as possible when transplanting them once they are established.

Aftercare

Keep watering to a minimum to maintain a well-aerated compost, which encourages bud development and reduces the likelihood of rotting. In fact there is probably no need to water at all if the root cuttings were initially planted in a reasonably moist compost and a humid environment is maintained.

Very often when the bud first develops, it produces a stem and green leaves but no root system. This will grow later from the base of the new stem. Even if the new roots do develop from the cutting, they too will not appear until after the stem and green leaves have grown. Do not water until the roots appear as the cutting is still liable to rot.

Place in a well-lit area once the stem appears. Harden off any young plants propagated in a warm (21°C/70°F) environment before planting out or potting up. Apply a liquid feed, according to the manufacturer's instructions.

3 Strike off the grit until level with rim. Label, and leave cuttings to develop.

4 Do not water until the roots have appeared. Then apply a liquid feed.

Tuberous roots

Some herbaceous perennial plants die back to a crown of buds each dormant season, and their roots are modified to store food. These specialized swollen roots are described as tuberous roots. They can be distinguished from modified stems by their structure and from root cuttings by their inability to produce adventitious buds on isolated roots and so grow a new plant.

There are two basic kinds of tuberous roots: those that develop annually, such as on dahlias, and those that are perennial and simply increase in size, such as on begonias.

Annual tuberous roots develop from lateral roots at the crown of the plant. During the growing season, certain of these develop as food stores, swelling up and producing a cluster of such roots. Each year the new shoot system develops at the expense of the food store in the tuberous roots, which eventually die and disintegrate.

Perennial tuberous roots are much simpler in their development. Usually the emerging radicle of the seedling begins to modify as a food storage organ, and this increases in size as and when food is available.

The division of tuberous roots is not a widely used technique as many plants with tuberous roots can be propagated more satisfactorily by stem, leaf and leaf-bud cuttings. Success depends on how well the roots are stored. Lift the plant at the end of the growing season. Clean the crowns and dust with sulphur. Wrap each plant in thick newspaper and place in a frost-free environment, below 5°C/42°F.

Just before the growing season, divide the tuberous roots into portions, each with at least one crown bud, from which the new stem will develop. Protect all cut surfaces from rotting by dusting with sulphur.

Then place the divisions in a warm (21°C/70°F), dry airy area for a couple of days so that the cut surfaces seal themselves quickly by developing a corky layer of tissue, to give added protection. Pot up the divisions in John Innes No. 1 compost if they are to be transplanted within a month or so; otherwise plant them in John Innes No. 2 compost. Label them; do not water. Place in a frost-free area. Set in the light once a shoot appears.

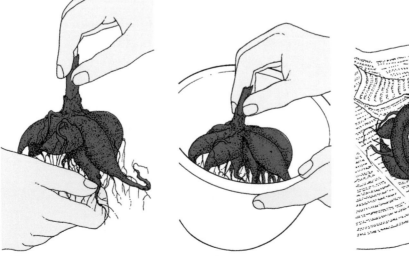

1 Lift a plant at the end of the growing season. Clean the crowns thoroughly.

2 Dust the entire crowns with a fungicidal powder. Lift on to some newspaper.

3 Wrap up the plant. Store in a frost-free place until the buds begin to swell.

PERENNIAL TUBEROUS ROOTS

Plants such as begonias have only one tuberous root, which is perennial and extends sideways each year.

ANNUAL TUBEROUS ROOTS

Some tuberous rooted plants such as dahlias produce annual storage roots that die and disintegrate after one season.

4 Divide the swollen roots into portions, each with at least one crown bud.

5 Dust all cut surfaces with fungicide. Leave in a warm, dry, airy place.

6 Pot up the cuttings once the cut surfaces have formed a corky layer.

Modified stems

A modified stem is an organ that stores food, which the plant can then use to survive its dormancy period. Also, it is often the means by which a plant can spread and produce new plants.

To be classed amongst this group a modified stem must exhibit all the characteristics of a stem. It must have a stem structure: that is it will have an apical growing point; the stem itself will carry leaves with buds in their axils; and the arrangement of the leaves will be spiral, alternate or opposite each other.

The stem is modified from the normal in that it may not necessarily be above ground, it often grows more or less horizontally, and it usually acts as a food storage organ.

There are basically only six kinds of modified stem, all with distinct habits and growth patterns. These are tubers, rhizomes, corms, bulbs, offsets and runners. They should not be confused with modified root systems, which do not possess all the features associated with a stem.

A tuber, for example, has "eyes", which have a cluster of buds and a leaf scar. These are the nodes of a typical stem. They are arranged in a spiral, alternately or opposite each other beginning from the apical bud on the end opposite the scar where the tuber was attached to the parent plant.

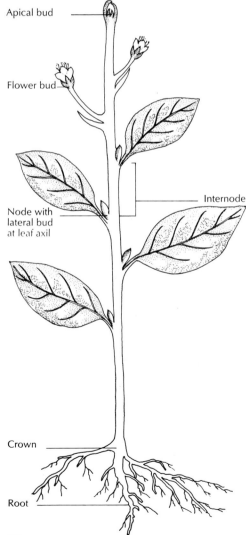

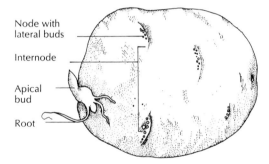

A rhizome is a specialized stem that grows horizontally just below the ground or on the surface. The stem appears segmented as it is composed of nodes and internodes, the leaves and flower bud developing from apical or lateral buds on the stem. Roots and a lateral shoot also grow from these points.

A corm is the swollen base of a solid stem which is surrounded by scale leaves. These are attached to the stem at distinct nodes which have a bud in each axil. At the apex of the corm is a bud that will develop into the leaves and the flowering shoot.

An offset is a lateral shoot that develops from a leaf axil at the crown of a plant. This lateral shoot usually has only one bud, which develops into a plantlet with its own root and growing point.

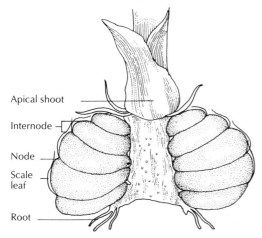

A bulb, on the other hand, has a short, fleshy, usually vertical stem which, at its apex, has a growing point enclosed by thick, fleshy scale leaves. New bulbs develop in the axil of these scale leaves. The flower bud and foliage leaves grow up from the centre of the bulb.

A runner also develops from the leaf axil at the crown of a plant. It grows horizontally and, at several of its nodes, forms new plantlets from the lateral buds. These new plants will develop a flowering point and also send out their own runners once they have established themselves.

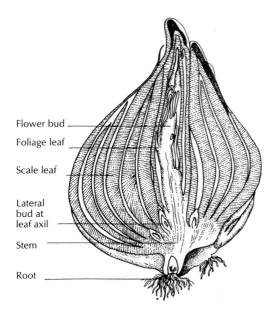

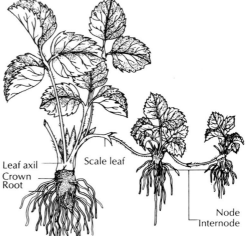

Frequently, modified stems are referred to as stolons, but this term has purposely been avoided in this book as it has a multiplicity of meanings.

Tubers

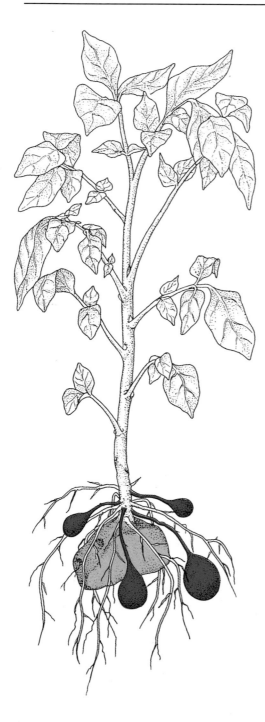

The name "tuber" in horticultural terms has been much misused and has on occasions covered almost any swollen food storage organ that is chiefly underground.

A tuber is, in fact, a swollen underground stem, modified as a food storage organ. It often is of roundish (usually terminating) growth and is normally annual. The leaves are scaly and membranous, and the axillary buds provide the following year's shoots.

Within these limitations the number of plants producing true tubers is not very great. The commonest is the potato. Those plants, such as Jerusalem artichokes, that produce opposite rather than spirally arranged buds, such as potatoes, often have a knobbly shape. Some water lilies (*Nymphaea* sp.) produce small tuber-like structures that develop from the main rootstock towards the end of the growing season.

Although potatoes are prolific producers of tubers, this is unusual. Normally, plants that develop tubers do so only in very small numbers.

TUBERCLES

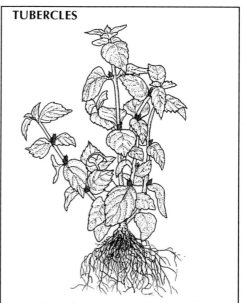

Just a few plants produce very small tubers from the axillary buds. These buds modify into tubers and eventually fall off the parent stem.

Some plants that produce tubers
Artichoke, Chinese
Artichoke, Jerusalem
Caladium
Nymphaea sp. (Water lilies)
Potato
Tropaeolum sp. – especially *T. tuberosum*

Tuberous pea

Some plants that develop tubercles
Achimenes
Begonia grandis ssp. *evansiana*
Dioscorea batatas

Any increase in numbers must be achieved artificially. As the tuber is normally an organ that allows a plant to survive its dormant period, the season to propagate tubers by artificial division should be just before growth would commence in the spring.

Cut the tuber with a sharp knife so that each piece has at least one good dominant bud or "eye". Then dust all the cut surfaces with a fungicide such as dry bordeaux powder to reduce the possibility of fungal infection. Stand the pieces on a wire tray and keep in a warm (21°C/70°F), dry environment, for example an airing cupboard, for a couple of days. A protective corky layer, which enhances their survival, will then develop.

These tuber "seeds" should not be kept in the dry for any longer than necessary, otherwise the tuber itself will desiccate. Therefore plant them out immediately in a hole twice their depth. They will then quickly produce roots and shoots and establish as a new plant. Label them clearly at all stages of propagation.

POTATOES

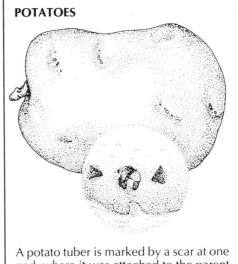

A potato tuber is marked by a scar at one end, where it was attached to the parent plant, and by "eyes", or nodes' placed spirally over its surface.

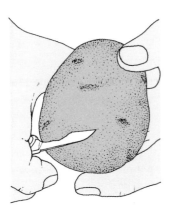

1 Cut a tuber into pieces with a sharp knife. Ensure each piece has at least one good "eye"

2 Dust all the cut surfaces with a fungicidal powder. Stand on a wire tray in a warm (21°C/70°F), dry place.

3 Plant the tuber pieces in the open ground as soon as a protective corky layer has formed. Label them clearly.

Rhizomes

A rhizome is a stem that grows laterally at about the soil surface, although in some plants it is underground. Normally, a rhizome stores food, but the degree to which it does this varies from species to species.

A rhizome is a perennial, and it is propagated artificially by division at an opportune season of the year: in most cases this is after flowering, when the rhizome is about to extend and produce new roots.

It has two ways of growing. In one, typified by the German or bearded iris, a terminal bud develops and flowers; the plant sends out extension growth through a lateral bud. The following season, this extension growth develops its own terminal bud, which flowers, and the plant continues to extend through its lateral buds. In the other way of growing, typified by mint and couch grass, its extension growth develops continuously from the terminal bud and occasionally from a lateral bud, which usually produces flower spikes.

In habit, rhizomes may also vary: "crown" rhizomes, such as asparagus, have virtually no extension growth annually and develop as a spreading crown, whereas other kinds of rhizomes, such as mint, couch grass and ground elder, make rapid and continuous growth and spread over large areas of ground in a relatively short time.

Perhaps the most prolific of garden plants to be propagated using rhizomes are bearded irises. The best time to divide their rhizomes and establish new plants is immediately after flowering, when the old root system dies down and a new root system begins to develop. Lift the clump of rhizomes with a fork and knock off as much soil as possible. Cut away and discard any old rhizomes, just leaving the current season's flushes of growth. Cut back their roots to 2–3 in, and shorten the leaf blades to reduce water loss before the new root systems develop. The prepared rhizomes are now ready for planting.

As a general rule, a rhizome should be replanted at the same depth as it was growing; for irises this is more or less in the surface soil. Usually, a rhizome has two rows of roots longitudinally on each side underneath. Therefore, when replanting, dig out two linear shallow trenches and place the roots in these. Firm back the soil over these roots and label the rhizome clearly. If necessary, settle in by watering.

1 Divide rhizomes after flowering, when the old root system dies down.

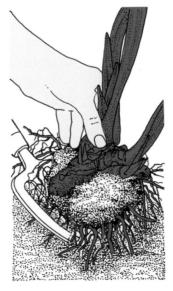

2 Lift a clump with a fork. Knock off as much soil as possible from the roots.

3 Cut away and discard any old rhizomes from the current season's growth.

Some plants that develop rhizomes
Arum lily (*Zantedeschia* sp.)
Asparagus
Begonia rex
Bird of paradise
  (*Strelitzia reginae*)
Canna

Couch grass
Ferns – some species
Ground elder
Iris, bearded
Lily of the valley
  (*Convallaria majalis*)
Mint
Mother-in-law's tongue

(*Sansevieria* sp.)
Peony
Rhubarb
Smilax
Solomon's seal (*Polygonatum*)
Thistle, creeping

CROWN RHIZOMES

Less easy to propagate are the "crown" rhizomes, such as peony and asparagus, which have what is traditionally referred to as herbaceous perennial rootstock. Fairly massive cuts are needed to divide such rootstock into suitable pieces for propagation: each piece requiring at least one well-developed bud.

Divide crown rhizomes in later winter before the buds enlarge and before the new season's root system begins to develop. Make a few relatively large divisions, unless quick bulking-up is essential, when it is better to plant small divisions in pots filled with John Innes No. 1 compost or similar (see page 23) in order to give them good conditions in which to become established. Dust the cut surfaces thoroughly with a fungicide to prevent bacterial and fungal rots. Leave them in a warm (21°C/70°F), dry atmosphere for the surface to dry out and develop the beginnings of a protective corky layer. Then plant out the divisions. Label them clearly.

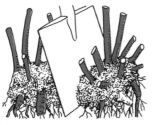

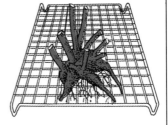

1 Divide crown rhizomes in later winter.

2 Dust the cut surfaces with talcum powder.

3 Plant out aftr two days in a warm, dry area.

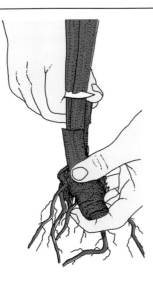

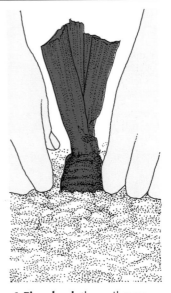

4 Shorten the leaf-blades and cut back the roots to 2–3 in.

5 Replant each rhizome on a ridge with its roots in trenches either side.

6 Firm back the soil over the roots. Label and settle by watering.

Corms

Corms look very similar to bulbs and are often confused with them. However, structurally they are very different. A corm consists of a stem that is swollen as a food store and that is shorter and broader than a bulb. The leaves of the stem are modified as thin, dry membranes that enclose the corm and protect it against injury and drying. Each leaf has a bud in its axil, the top of the stem usually develops as a flowering stem, and the roots are produced from the corm's base, which is often concave. In some kinds of corm, several buds at the top of the stem may grow out and flower.

Each year a new corm develops around the base of each stem. Increase, therefore, is

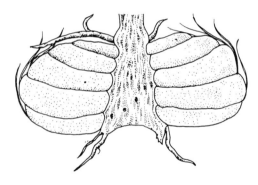

directly related to the number of stems produced by a corm. Normally, most plants developing corms will propagate naturally to give a sufficient increase, but should it be necessary to bulk up supplies more quickly then an artificial technique should be used.

Always buy corms from a reputable specialized grower, because it is vital to propagate from disease-free corms.

Cut a large, healthy corm into several pieces just prior to the season for planting and ensure each piece has at least one bud. Dust the cut surfaces with a fungicidal powder such as dry bordeaux or sulphur, in order to reduce the risk of rotting. Set the pieces on a wire cake-tray and place in a warm, dry environment, such as an airing cupboard, for 48 hours. This will cause the cut surfaces to seal. Then plant singly in a pot or in the ground; and label clearly.

If the corm is too small to cut up satisfactorily, then the lateral buds can be induced to develop more readily by removing the main stem either by snapping it off or digging it out with a knife. Then dust the cut surface with fungicidal powder and plant out the corm in the ground. During the growing season it will produce several shoots, which will eventually become new plants.

CORMELS

Cormels are miniature corms that are produced as offsets between the new corm and the old disintegrating corm. The quantity produced is a widely varying feature — gladiolus developing up to about 50 cormels.

The level of cormel production will be influenced by the depth at which the main corm is planted; the deeper the corm is in the ground, the more cormels produced.

Collect the cormels when the corm is lifted from the ground before winter and store them below 5°C/41°F in a dry, frost-free environment with air circulating round them. Soak any cormels that become dry in tepid water for 24 hours before planting the following season. Plant them outdoors close together and label. They will normally take two years to reach flowering size.

Some plants that develop corms
Crocosmia
Crocus
Gladiolus
Ixia
Ixiolirion
Montbretia
Tritonia

Propagating by division

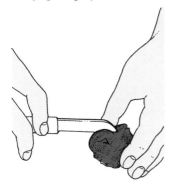

1 Cut a corm into several pieces, each with at least one bud, just before planting in autumn.

2 Dust the cut surfaces with talcum powder. Set pieces on a wire tray. Leave in a warm, dry place.

3 Plant each piece in a pot or in the open ground once it has developed a corky layer. Then label.

Propagating by inducing lateral buds

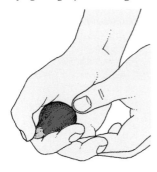
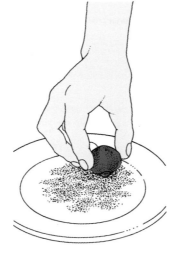
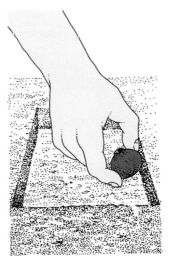

1 Lift the corm in autumn. Snap off the main stem or dig it out using a pointed knife.

2 Dust all of the cut surfaces with talcum powder in order to dry them.

3 Dig a hole twice the depth of the corm in the open ground. Plant it out immediately and label.

Bulbs

Bulbs are modified stems in which the scale leaves are modified for food storage. There are two kinds of bulbs, tunicate and scaly, and they differ in the development of their scale leaves.

Tunicate bulbs

Tunicate bulbs, such as daffodils and tulips (see below), have fleshy, very broad scale leaves, more or less surrounding the previous leaf so that the leaves make nearly complete concentric rings around the growing point. Each scale leaf has an axillary bud. The outer scale leaves become dry and membranous and give the bulb good protection against drying and injury. The roots of a tunicate bulb develop at the beginning of its growing season on the outside edge of the basal plate.

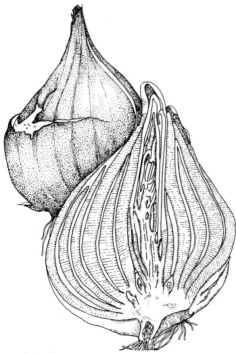

Scaly bulbs

Scaly bulbs, such as fritillaries, *Nomocharis* and lilies (see below), do not develop a dry, membranous covering and are very much more susceptible to drying than tunicate bulbs. The leaves are normally scaly and very

fleshy, but are often relatively narrow. The roots are produced in midsummer or later, and persist through to the following year.

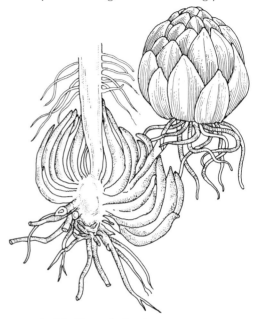

Natural bulb reproduction

Bulbs propagate naturally by division and for some this is the only way they can be propagated. In the annual growth cycle the apical bud develops and produces a new bulb during the growing season. If an axillary bud develops into an active growing point, then this also develops as a bulb that may take a year or two of further growth before it separates from its original parent (see below) and

Some lilies that produce bulbils as a natural growth process
Lilium bulbiliferum
L. sulphureum
L. sargentiae
L. lancifolium
and its hybrids

eventually starts flowering. In some plants, notably tulips and bulbous irises (see below) the original bulb disintegrates after flowering, leaving a cluster of small bulbs as well as a new flowering bulb. In autumn, pull these apart and plant out at twice their own depth.

Bulblets are offsets that develop on some lilies, such as *Lilium longiflorum* and *L. bulbiferum*, just below ground level either above or below the main bulb.

In late summer, gently detach any bulblets and plant the bulb and bulblets straight into the ground at twice their own depth.

Unfortunately, only a few species naturally grow bulblets in any quantity, although some other important species, such as *L. auratum*, *L. speciosum* and *L. lancifolium*, do produce a small number.

Bulbils are tiny bulbs that grow in the leaf axils of stems, which are above ground, of certain species of lily such as *L. lancifolium*.

After flowering time, collect the bulbils off the plant as they mature. Set them 1 in apart in a pot filled with John Innes No. 1 compost or similar (see page 23). Cover with grit and place in a cold frame. In autumn of the following year, transplant the bulbils into the ground.

Bulblets and bulbils

Bulblets

Bulblets are tiny bulbs that develop below ground on some bulbs.

Plants that produce bulblets can be artificially induced to increase their bulblet production. This is done by removing the flower stem, and burying it until bulblets develop in the leaf axils.

Dig a trench 6 in deep; slope one side up to ground level. Pinch out any buds or flowers on the stem and then twist it out of the bulb, which should remain in the ground. Lay the stem in the trench along the slope, leaving part of it sticking out of the ground. Spray the stem with a liquid fungicide to prevent disease. Then fill the trench with sand or a light compost and label.

By autumn, bulblets will have developed in the leaf axils at the lower end of the stem. These can be detached and planted straight into the ground at twice their own depth or be left *in situ* for a year.

This is a surprisingly easy method of producing bulblets, and the only difficulty that is likely to arise is the potential rotting of the stem before the bulblets are produced.

Bulbils

Bulbils are tiny bulbs that grow on a stem above ground. This natural process is an extremely prolific way of increasing stock.

A number of lily species such as *Lilium candidum* can be artificially induced to produce bulbils by disbudding the plant just before flowering. Bulbils will develop in the leaf axils during the remainder of the growing season. Collect them as they mature.

Fill a pot to the rim with John Innes No. 1 compost or similar (see page 23). With a presser board, strike off the excess compost and then firm gently to within ³/₈ in of the pot rim. Set the bulbils 1 in apart on the compost surface and gently press them so they make contact with the compost. Cover generously with grit. Strike off the grit level with the rim of the pot. Label and place in a cold frame. Leave for at least twelve months, until autumn of the following year. Then transplant into the open ground.

Certain other plants can also be artificially induced to produce bulbils; some of the ornamental onions will develop them in the flower heads.

Bulblet propagation

1 Pinch out any buds or flowers on a suitable plant.

2 Twist the stem out of the bulb, leaving the bulb in the ground.

3 Lay two-thirds of the stem along the sloping side of a trench 6 in deep.

Some bulbs that produce bulblets
Lilium auratum
L. bulbiferum
L. canadense
L. longiflorum
L. pardalinum

L. speciosum
L. lancifolium

**Some bulbs that produce
bulbils as a result of debudding**
Lilium candidum
L. chalcedonicum

L. dauricum
L. × hollandicum
L. leichtlinii
L. × maculatum
L. × testaceum
Ornamental onions
— some species

Bulbil propagation

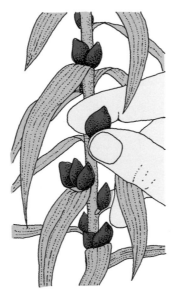

1 Disbud a suitable plant stem just before it begins to flower.

2 Pick off any bulbils as they mature in the leaf axils on the stem.

3 Set them 1 in apart in a compost-filled pot. Cover with grit and label.

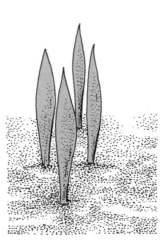

4 Fill the trench with sand or a light compost. Label clearly.

5 Leave during the summer. Detach the bulblets in autumn and replant.

Plants that produce bulblets include:
Allium, Cardiocrinum, Chionodoxa, Erythronium, Fritillaria, Galanthus, Hyacinthus, Leucojum, Lilium, Muscari, Narcissus, Nerine, Scilla, Sternbergia and *Tulipa*.

Plants that produce bulbils include:
Summer-flowering species of *Lilium* and *Allium*, and culinary members of the *Allium* genus such as *A. cepa* and *A. sativum*.

Bulb scaling

To increase a particular variety of bulb rapidly it is necessary to use artificial techniques, because the natural rate of increase, although steady, is generally slow.

One technique is to take leaf cuttings, although, with bulbs, these leaves are modified in the form of bulb scales and the technique is known as bulb scaling.

Scaly bulbs, such as lilies and fritillaries, have relatively small, narrow scale leaves which can readily be pulled off the basal plate of the bulbs. When they can best be propagated depends on the availability of bulbs. Ideally, it is easier to deal with fresh bulbs, and these are normally available in October/November. Imported bulbs are not usually available until January/March, and these will invariably have wilted slightly, which makes them less than ideal as propagating material.

Take cuttings from a fresh bulb in a turgid condition by pressing the scale leaves outwards until they snap off close to the basal plate of the bulb. With a flaccid bulb, cut the scale leaves as close to the basal plate as possible, using a sharp knife.

To propagate from one bulb, cut off only a few scale leaves from the outside of the bulb. Some bulbs, for example lilies, do not like being lifted and replanted; therefore dig a hole around the bulb, break off some scales and replace the soil around the bulb.

Any scale leaves will carry potential rotting agents and so they must be protected. Place the scales in a polythene bag filled with a fungicidal powder such as sulphur dust or dry bordeaux powder. Shake the bag vigorously until the scale leaves are completely covered with a thin film of fungicide.

Mix the scale leaves with four times their volume of an extending medium, which will allow them to develop in a damp, well-aerated environment. Various materials, such as damp vermiculite, can be used very successfully, but, provided that it is adequately sterile, a mixture of equal proportions of damp peat and grit is just as suitable. Place the mixture and a label in a polythene bag. Blow into the bag; when fully expanded, tie its neck. Store in a warm (21°C/70°F) place, such as an airing cupboard, so that new plantlets can develop.

The rate at which regeneration occurs is related chiefly to temperature, but is also a varietal characteristic – some being quicker to propagate than others. Generally activity may be expected in about six to eight weeks.

The new plant will initially take the form of a bulblet and will develop on the broken surface at the base of the scale leaf.

4 Mix the scale leaves with four times their volume of damp peat and grit. Place in a polythene bag.

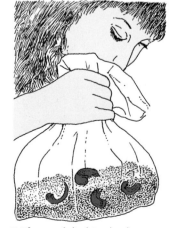

5 Place a label in the bag. Blow into it. Then tie round its neck. Store in an airing cupboard.

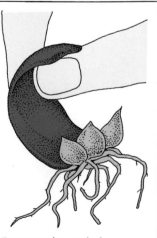

6 Remove the scale leaves from the polythene bag as soon as bulblets appear on the broken basal surfaces.

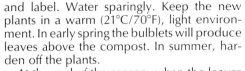

As soon as these bulblets appear, remove the scale leaves from their bag and plant them vertically in John Innes No. 1 potting compost or equivalent (see page 23). They can either be planted singly in a 3–3½ in pot or lined out into a deep tray, depending on numbers. Ensure the tips of the scales are just visible above the compost. Cover with grit and label. Water sparingly. Keep the new plants in a warm (21°C/70°F), light environment. In early spring the bulblets will produce leaves above the compost. In summer, harden off the plants.

At the end of the season, when the leaves have died down, lift and separate the new young bulbs. Replant immediately.

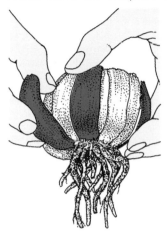

1 Remove a scale leaf from the bulb by pressing it outwards or cutting it close to the basal plate.

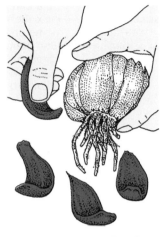

2 Snap or cut off only a few more scale leaves from the outside of the bulb.

3 Place the scale leaves in a bag filled with sulphur dust. Close bag and shake vigorously.

7 Plant each scale leaf with its tip just visible. Cover with grit and label. Place in a warm, light area.

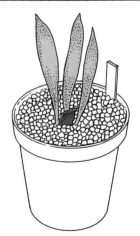

8 Harden off the plants in summer after the bulblets have produced leaves above the compost.

9 Separate new bulblets from their scale leaf once their leaves have died down. Replant and grow on.

95

Scooping and scoring bulbs

Some tunicate bulbs, such as hyacinth, grow larger each year and propagate very slowly by natural division of the bulb. They therefore have to be propagated artificially if a significant increase in numbers is required.

The scale leaves of tunicate bulbs are large and encircle the bulb; they are not as readily removed from the basal plate as the leaves of scaly bulbs. It is therefore necessary to leave their cut scale leaves *in situ* while inducing them to produce plantlets. There are two ways of doing this: scooping and scoring.

Scooping

This is carried out towards the end of the bulb's dormant season. To scoop the bulb successfully, and with minimum damage, requires a special tool: an old teaspoon with one sharpened edge is excellent. Use it to cut out the basal plate in one scooping movement, leaving the rest of the bulb undisturbed and the cut surfaces of all the scale leaves exposed. Although it is possible to do this with a knife, it is inadvisable as the centre of the bulb may become macerated and subject to rotting. Once the basal plate has been removed, dust the cut scale leaf surfaces liberally with a fungicidal powder to minimize potential rotting.

Set the bulb upside down, with the scale leaf bases exposed, on a wire tray or in a tray containing dry sand. Place in a temperature of at least 21°C/70°F to encourage calluses to form on the scale leaf bases and so further combat any chance of infection. At the same time, the bulb should be kept as dry as possible, but ensure that the scale leaves do not desiccate. An airing cupboard is probably a suitable environment, but dampen the sand occasionally.

In about two to three months the new bulblets will develop on the cut surfaces of the scale leaves. Plant the bulb in a pot, still placing it upside down so that the bulblets are just below the surface of the compost. Label clearly. Harden off and then leave in a cold frame.

In spring, the bulblets will grow and produce leaves and the old bulb will gradually disintegrate. At the end of the season lift and separate the bulblets, and replant. They will normally take a further three or four years before flowering size is reached.

Scoring

Flowering bulbs can be produced in a shorter time by using a similar technique called scoring, which makes fewer, larger bulblets as there are less cut scale leaf surfaces.

Scooping the bulbs

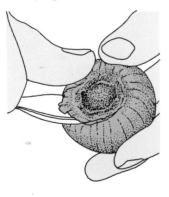

1 Sharpen one edge of an old teaspoon. Use it to remove the basal plate.

2 Ensure the base of every scale leaf is removed.

3 Dust the cut scale leaf surface with sulphur. Set upside down on a tray.

The process follows exactly the same pattern as for scooping except that, instead of cutting away the basal plate of the bulb, the basal plate is scored with a sharp knife. Cut through the basal plate of a tunicate bulb until the scale leaves are scored to a depth of about ¼ in. Make four equally spaced scores if the bulb is large; on smaller bulbs two scores at right angles will suffice.

Then place the scored bulb in a warm (21°C/70°F), dry environment for a day; this will cause the cuts to open out. Dust the cut surface with a fungicide such as sulphur dust or dry bordeaux powder. Subsequently, the treatment is the same as for scooped bulbs.

The bulblets produced in this way usually only require a further two or three years to reach flowering size.

Scoring bulbs

1 Make two scores at right angles to each other on the basal plate of the bulb.

2 Place in a warm, dry area until the bulb case opens out. Dust with sulphur,

3 Place on a tray. Store in an airing cupboard until the bulblets develop.

Growing on bulblets

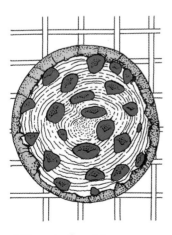

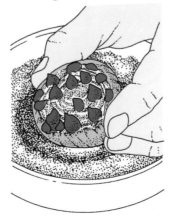

4 Store in the airing cupboard until bulblets appear on the cut surfaces.

5 Plant the bulb upside down with the bulblets just below the compost surface.

6 Lift and separate the bulblets at the end of the season. Replant at once.

Division 1

Herbaceous plants with fibrous crowns

1 Lift the plant that is to be divided directly after it has flowered.

2 Shake off as much soil as possible.

3 Wash the crown and its roots in a bucket, or hose it clean.

4 Shorten all tall stems above the ground to minimize water loss.

5 Break off a piece with at least one good "eye" from the edge of the crown.

6 Divide any intractable pieces with an old carving knife or similar blade.

Some herbaceous plants with fibrous crowns
Achillea
Alchemilla
Alyssum
Aster
Aubrieta
Caltha
Campanula
Chrysanthemum
Doronicum
Erigeron
Gentiana
Geranium
Helenium
Hemerocallis
Lupin
Lythrum
Mimulus
Monarda
Polemonium
Prunella
Pyrethrum
Raoulia
Rudbeckia
Sagina
Scabious
Tiarella
Trollius
Veronica

Dividing a plant is a common way to propagate many herbaceous perennials, and it is also used to rejuvenate favourite plants and keep them in a vigorous condition. Propagation by division is also successful with shrubs, such as sumach, that produce suckers; with semi-woody perennials, such as New Zealand flax, that produce a crown of offset shoots; and with most plants with modified stems, such as bearded iris (see pages 82–97).

Herbaceous plants with fibrous crowns

The commonest method of propagating plants by division is that used for herbaceous perennials, such as chrysanthemums, with fibrous roots and a relatively loose crown. Normally, the central part of the crown becomes woody over the course of two or three years. As this woody area does not produce many shoots and generally loses vigour, it is discarded and the remainder of the clump is divided into suitable-sized portions for planting out and re-establishing a new crown.

The only variable feature of this form of propagation is the time at which division is carried out. As a general rule, the most opportune time to divide such plants is directly after flowering, as this is when the new vegetative shoots are being produced and the new root system is developing. In very late-flowering subjects, this would be the following spring.

Lift the parent plant and shake off as much soil as possible. Then wash the crown in a bucket, or hose it clean of any residual soil. The plant can be divided without this preliminary preparation, but it is much easier to deal with clean plants, especially if the soil is wet and muddy. Shorten any tall stems above the ground to prevent unnecessary water loss, especially if the division takes place in summer. Break off a piece with at least one good "eye" from the periphery of the crown, where the young shoots are generally produced; avoid the central woody crown, which is of no value and should be discarded. If the piece proves rather intractable to remove, cut it off, using an old carving knife or similar blade. Plant out the new clump as quickly as possible to the same depth as it was growing previously. When replanted it should be labelled and watered in – indeed "puddled" in would be more appropriate.

7 Make a hole and replant the new clump at once. Firm the soil and label.

8 Water very thoroughly, using a watering can with a spray attachment.

9 Keep the new clump free of weeds.

Division 2

Herbaceous plants with fleshy crowns

Many herbaceous plants, such as hostas, develop a compact, fleshy crown that is not easy to pull apart.

. The best way to propagate these plants is by division towards the end of their dormant season, when buds will begin to shoot, indicating the most vigorous areas.

Lift the parent plant and shake off as much soil as possible. Wash the crown thoroughly. With a convenient-sized knife, cut the crown into pieces. The size of divisions will depend on preference, but must include at least one developed shoot. Avoid latent buds, which do not always develop satisfactorily. Dust the cut surfaces with dry bordeaux powder to reduce chances of fungal rots. Do not allow the divisions, especially from really fleshy rooted plants such as hostas, to dry excessively, Therefore replant the divisions either in the ground or in a pot as quickly as is feasible. Label them clearly.

Naturally dividing alpines

There are a number of alpine plants, such as campanulas, which lend themselves to propagation by division because their crowns separate naturally into individual new plantlets each season.

After flowering, or in the spring if the plant flowers in the autumn, as does *Gentiana sinoornata*, lift the plant and tease apart the divisions. Replant as soon as possible. Label and water well.

This is a very simple but effective system of increasing plants. If the crowns are lifted and divided fairly frequently the rate of increase can be quite dramatic. However, plants left *in situ* for a long period tend to produce only a few large divisions.

Semi-woody herbaceous plants

Some perennials with upright, sword-like leaves, for example *Phormium*, increase in size by producing a sort of offset that develops into a large crown of individual shoots, each with its own root system.

To propagate these plants, it is best to divide them in the spring, although it can be done at any time of the year. Lift them and shake out the soil, if necessary hosing or washing the crown clean. Pull the various pieces apart. Cut the clump with a spade or hatchet if it is hard and woody in the middle,

Herbaceous plants with fleshy crowns

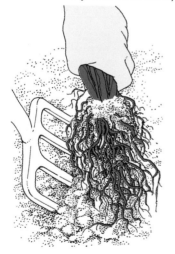  

1 Lift the plant to be divided towards the end of its dormant season.

2 Wash the crown well. Cut off a piece with at least one developed bud.

3 Dust the cut surfaces with talcum powder. Then replant immediately.

Some herbaceous plants with fleshy crowns	Campanula	Some woody shrubs
fleshy crowns	*Primula* – European species	*Aronia*
Astilbe	*Thalictrum*	Blackthorn
Hosta		Lilac
	Some semi-woody	
Some alpine plants dividing	**herbaceous plants**	
naturally	*Cortaderia*	
Aubretia	*Phormium*	
Autumn gentian	*Yucca*	

and replant the divisions fairly quickly to avoid the roots dying.

Woody shrubs

There are a few woody shrubs, such as blackthorn, that produce suckers, which then develop into individual clumps of stems. Lift these plants in the dormant season and wash thoroughly. Divide the clump of stems into convenient-sized portions. Normally the main core of the clump will be woody and will carry few roots. This will be of little value for propagation purposes, so take the new pieces from the younger, more vigorous growth on the outside of the clump.

Cut back the branches fairly drastically to reduce water loss, as the buds will break in the spring before sufficient roots have been produced. Replant the divisions back in the ground as soon as possible and label.

Lift relatively isolated suckers individually in the dormant season and use to establish new plants.

If the plant has been grafted – as may happen with Japanese quinces – then it is the rootstock, not the cultivated variety, that is divided.

IRISHMAN'S CUTTINGS

Plants, such as Michaelmas daisies, that produce particularly loose crowns can be propagated by separating off single stems on the periphery of the crown so that each has an adventitious root system. These single stem portions are described as "Irishman's cuttings", and they should be planted at once.

Woody shrubs with suckers

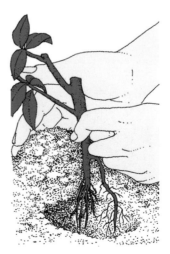

1 Remove isolated suckers from woody shrubs during the dormant season.

2 Cut back part of the roots and top growth.

3 Replant immediately; label and water in well.

101

Offsets/Runners

Offsets

An offset is a plantlet that has developed laterally on a stem either above or below ground: the stem arises from a crown bud and usually carries no other buds.

Most plants, such as sempervivums, that produce offsets first grow a miniature plant with only minimal roots. A root system will not fully develop until late in the growing season.

To speed up this process of propagation, pull away the offset from its parent, usually in spring. Either plant out in the garden or pot it up, using a cuttings compost with added grit, which will drain freely and so ensure good root development.

Where offset development is poor, it is possible to stimulate offsets by removing the plant's growing tip. This has the same effect as removing a terminal bud.

Although the term offset is normally used for a plantlet that is separated from its parent during the growing season, it is also used to describe slower-developing shoots that are produced by mainly monocotyledonous plants such as yuccas. Eventually these shoots should be sufficiently mature to be separated from the parent and planted on.

A pineapple plant produces offsets that can be used for propagation once its fruit is nearly mature. These offsets are variously described as slips, ratoons or suckers. Cut them off the plant close to the crown and plant out.

Many corm and bulb plants, such as fritillaries, each year produce miniature reproductions of themselves from the bases of the newly developing corms or bulbs. These are known as cormels (see page 88) and bulbils and bulblets (see page 92).

Runners

A runner is a more or less horizontal stem that rises from a crown bud and creeps overground. The leaves are normally scale-like, and rooting may occur at the nodes. The lateral buds develop as new plants, and eventually the stem of the runner deteriorates, leaving a new isolated plant. The classic example of this kind of natural vegetative reproduction occurs in the strawberry plant.

New plantlets usually root down and produce new plants quite successfully. However, unless controlled, a mat of new plants tends to develop, and these are not easily lifted and separated without damage, so thin out the runners regularly.

1 Thin out some runners in early summer to encourage strong growth.

2 Fill a pot with John Innes No. 1 compost. Firm to within ⅜ in of the rim.

3 Dig a large hole in the ground beneath a plantlet. Set the pot in the hole.

Some plants developing offsets
Agave
Crassula
Echeveria
New Zealand flax (Phormium)
Pineapple
Sempervivum
Yucca

Some plants developing runners
Ajuga reptans
Geum reptans
Grass – some species
Potentilla, herbaceous species
Saxifraga sarmentosa
Strawberry

1 Pull an offset away from the parent plant, preferably in spring.

2 Plant the offset in a pot filled with a cuttings compost with added grit.

Cut off pineapple slips close to the crown and plant out individually.

To produce large, well-established individual plants, dig in plenty of good compost for rooting; in early summer, thin out some of the runners, and pin down the rest into the compost, evenly radiating them around the plant. This method will induce early rooting, but the rate of development will not be quite as fast as for pot-grown runners.

These are obtained by placing a pot under each runner.

Dig a hole in the ground beneath a developing plantlet. Set a pot containing good compost into it and then push back the soil to keep the pot in position. Pin down the runner, using wire bent in the shape of a staple, so that the plantlet will root in the pot.

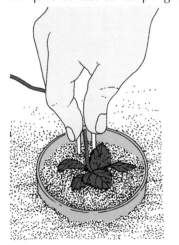

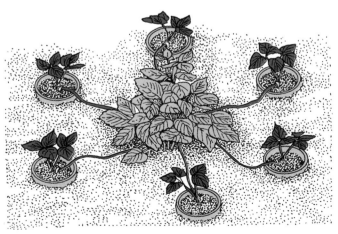

4 Pin down the plantlet in the middle of the pot, using a wire staple.

5 Pin down any other plantlets in a star-shaped pattern round the parent plant. Sever their connecting stems once they are fully established.

Stems

The propagation of plants from stems is one of the most widely practised techniques of vegetative propagation.

The technique involves initiating, developing and establishing a root system on a stem either before or after that stem has been removed from the parent plant, and this can be done in two ways: by layering and by stem cuttings.

With layering, the stem is encouraged to produce roots before it is severed from the parent plant. The main problem is to establish the rooted layer after it has been removed from the parent.

With stem cuttings, the stem is removed from the parent plant before it is encouraged to develop roots. With this method, it is difficult to keep the cutting alive until its roots have grown and established. Its advantage is that, usually, it takes up much less space than propagating by layering.

Always propagate from stems that have a high capacity to produce roots. The gardener must learn to judge this capacity in plants, as it has a significant affect on the eventual success of the layer or cutting.

The age of the parent plant, as well as that of the actual stem to be propagated, both have major influences on the capacity of a stem to produce roots.

A seedling that has recently germinated cannot immediately flower and produce

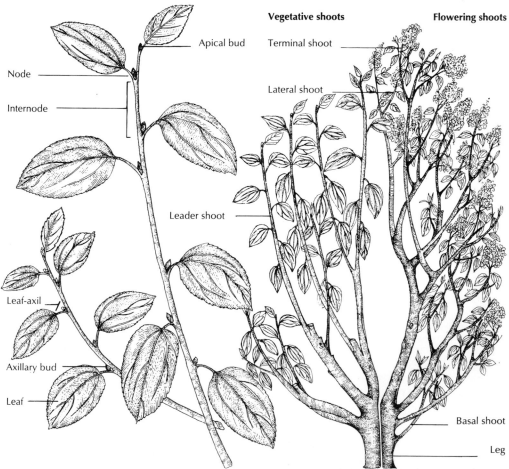

Apical bud

Node

Internode

Leader shoot

Leaf-axil

Axillary bud

Leaf

Vegetative shoots

Terminal shoot

Lateral shoot

Flowering shoots

Basal shoot

Leg

seeds, as it is not sexually mature. Therefore it is described as being juvenile, and the only way it can reproduce is asexually, from vegetative parts. Many juvenile plants have a very high capacity to regenerate vegetatively. The juvenile stage is not transient but fixed in some ornamental plants, such as some varieties of ivy (*Hedera helix*) and the Lawson's cypress varieties 'Ellwoodii' and 'Fletcheri', and these have a greater capacity than their adult varieties to produce roots on stems.

Juvenile plants, however, do not have much value in the garden as they do not flower and produce seed. Most of them will eventually mature into adult plants, when they will flower. In this condition a plant is sexually mature, capable of regenerating by producing seeds, and so it does not need to reproduce vegetatively.

A plant's capacity to regenerate from vegetative parts declines with age. Thus the gardener wishing to propagate vegetatively from a mature plant encounters a problem. This can be overcome by growing a more mature plant that does not produce flowers, so that its capacity to regenerate vegetatively may be increased, and thus the rooting capacity of its stems.

To prevent the plant, or at least the stems used for propagation, from flowering, prune them rigorously so that strong, vegetative (that is non-flowering) shoots are produced. The harder the stems are pruned, the faster they will grow and the more they will produce roots. The first flush of growth on a stem always has the greatest capacity to develop roots. Really fast growth is achieved by a combination of hard pruning and forcing the plant in a high temperature, at least 16°C/60°F. Under these conditions, roots may be induced on otherwise quite intractable stems.

The capacity of a stem to root will also be dependent on the age of the parent plant. The older the plant, the less it will be able to produce roots on stems, even if the stem to be propagated has been severely pruned. If the plant has been grown from a seed by the gardener, then he will know how old it is, and if he can successfully propagate from it.

If the plant to be propagated has been produced by a vegetative technique quite recently, it should root from its stems quite readily, but this is not always so. Many plants, however recently they were vegetatively propagated, may in fact be quite old! If the original plant was grown from seed and displayed attractive ornamental characteristics, it would always have been propagated vegetatively to maintain those characteristics in subsequent generations, but its inherent capacity to produce roots on stems will have declined, despite a temporary resurgence after pruning.

Thus new (that is, young) varieties are generally easier to propagate from stems than older varieties. This is particularly evident with plants such as deciduous azaleas, in which the old Ghent varieties are over 140 years old, the newer Ilam varieties are only 20 years old, and the Exbury varieties are 40 years old. The last two groups are relatively easy to propagate from stems.

Before propagating from stems, it is also necessary to consider whether there is a best time of year in which to propagate a particular plant by a particular technique, as it is pointless to do so in the wrong season. The capacity of many plants to produce roots on stems varies considerably during the different seasons, whereas in others there is no distinct seasonal variation. The index at the back of this book will tell the gardener when, where and how to propagate a particular plant.

STRUCTURE OF A WOODY STEM

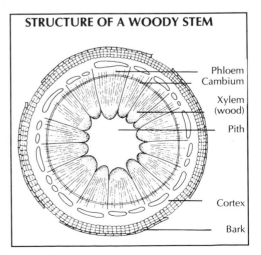

Phloem
Cambium
Xylem (wood)
Pith
Cortex
Bark

Layering/Simple layering 1

Layering is one of the oldest techniques used by gardeners to propagate woody plants. A stem is encouraged to develop roots before it is removed from the parent plant. This method is easy as it avoids any complicated environmental control to ensure the stem survives while the root system is developing. The early gardeners had probably observed this process when it occurred naturally under certain conditions. To do it artificially they just repeated the conditions and modified the technique.

As with any method of propagation, it is necessary to consider both the stem from which the layer is to be made and what soil is suitable to encourage rooting.

The condition of the stem will have a significant affect on the eventual success of propagating by layering. For greater success, the gardener should develop vigorous, rapidly grown stems with their high capacity to produce roots. Certain methods of layering, such as stooling, encourage this capacity in stems as part of the normal system of management, but in others, such as simple layering, it is necessary to prepare the stems by pruning the parent plant well before propagation can take place. When pruning bear in mind that, with most methods of layering, the branches to be layered will have to be brought down to ground level.

The soil around the parent plant will also require preparation so that it will induce the stem to form roots. Rooting will be encouraged primarily by the exclusion of light, but also by the availability of oxygen and the presence of sufficient moisture and warmth.

The exclusion of light from the stem, that is blanching it, is extremely important when encouraging roots to start growing. The sooner light is excluded from the stem, the more effective is the response. So, the earlier a stem is buried or earthed up the more likely it is to root. This effect cannot be overemphasized, as lateness in earthing up is probably the commonest reason why a stem fails to produce roots, provided that it is basically vigorous and capable of producing roots.

The soil for layering must have a good water-holding capacity, good aeration and adequate drainage. Thus, especially if plants are to be layered in an ordinary garden border as opposed to a purpose-developed layer bed, the soil must be dug deeply to provide good drainage and so reduce the chances of waterlogging. It can be further lightened and improved by the addition of peat and/or grit, depending on the heaviness of the soil.

Warmth will improve rooting, so ensure the layered stem and soil are placed where they will receive adequate sunlight. However, warmth will be effective only if the soil is moist, so water the layered stems during dry periods.

In most methods of layering, the soil should be carefully forked away from the layer once it has rooted, so it can be lifted. Do not allow the roots to dry out, otherwise they will die.

Layers that may have difficulty rooting successfully should be well established before they are lifted. To encourage this, sever the newly rooted layers from the parent plant about three to four weeks prior to lifting and replanting.

This weaning can be further enhanced by pruning the stems so that there is a greater balance of root to stem.

Simple layering

Simple layering is perhaps the easiest and most effective method of layering a wide variety of woody plants, and it is a technique that can be carried out with minimum disturbance to the parent plant.

A stem is buried in the soil behind its tip so that roots are induced in this area. Once the root system is established, the stem can be severed from the parent. The roots are encouraged to develop because the plant foods and hormones are restricted where the stem is buried. To be successful, the gardener must use stems that have a high capacity to produce roots, and they should also be near ground level.

Twelve or more months before layering, rigorously prune a low branch or branches on the parent plant so that young, rapidly grown shoots are produced. These will be more amenable to the bending and manipulation involved in the actual layering operation, and because of their rapid growth will have the required capacity to produce roots.

Layering is normally done in late winter/early spring, as soon as the soil can be worked down to a tilth.

As the process is likely to be carried out in the garden and not in a purposely-prepared nursery bed, it is important to prepare the soil both thoroughly and effectively. Dig as deeply as possible. Then add peat and grit in sufficient quantities to convert the existing soil into a rooting medium with good water-holding capacity, good aeration and adequate drainage.

Trim a rapidly grown stem of side-shoots and leaves for about 4–24 in behind the tip. Pull the stem to ground level and mark its position on the soil 9 in behind its tip. Dig a trench from that point, with one straight side about 4–6 in deep and the other sloping up to ground level near the parent plant.

The secret of inducing root formation is to restrict movement of food and hormones in the tissues of the stem: this is usually achieved by bending the stem at least at a right angle.

However, in plants that are particularly difficult to root, the stem should first be girdled by cutting into the stem with a knife or by binding the stem tightly with a piece of copper wire at the bend.

Bend the stem at right angles 9 in behind its tip and set it in the trench against the straight side with its tip exposed above the trench. If the stem is whippy, peg it down with heavy wire staples. Bury the stem with soil, firm in and water well.

Keep the soil reasonably moist, especially in dry periods. Rooting will normally occur during the growing season.

In autumn, sever the layered stem from the parent plant so that the new plant can accustom itself to an independent existence.

About three to four weeks later, cut off the growing tip to encourage the roots to establish. Pot up or plant out the layer and label it. If rooting is not well advanced by autumn, leave the layer to establish for a further year before lifting and transplanting it.

WAYS TO GIRDLE A STEM

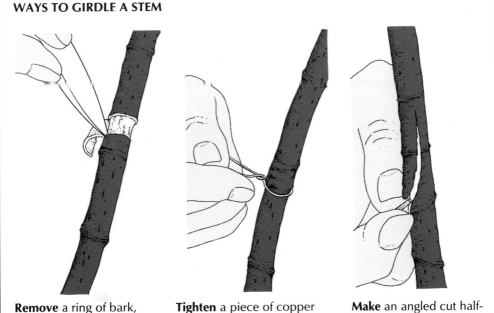

Remove a ring of bark, about ⅓ in wide, from round the stem, using a sharp knife.

Tighten a piece of copper wire round the stem. Twist it finger tight.

Make an angled cut halfway through the stem. Keep the cut surfaces apart with a matchstick.

Simple layering 2

1 Prune some low branches off the parent plant during the dormant season to induce vigorous growth.

2 Cultivate the soil well round the plant during the following late winter/early spring. Add peat and grit.

3 Trim the leaves and side-shoots off a young, vigorous stem for about 4–24 in behind the growing tip.

7 Return the soil to the trench to bury the stem. Firm it in well.

8 Water well, using a watering can with a coarse rose. Keep the soil moist, especially in dry periods.

9 Sever the layered stem from its parent plant in autumn.

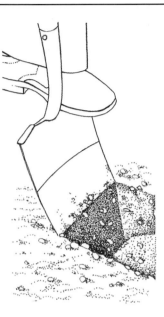

4 Bring the stem down to ground level and mark its position on the soil 9 in behind its tip.

5 Dig a trench from that point. Make one straight side 4–6 in deep. Slope the other towards the plant.

6 Bend the stem at right angles 9 in behind its tip. Peg it down in the trench against the straight side.

10 Cut off the growing tip from the rooted layer about three to four weeks later.

11 Lift the layered stem if its roots are well advanced. Otherwise, leave for a further year.

12 Replant either in the open ground or in a pot and label. Leave to establish.

Air layering

Air layering is one of the oldest artificial techniques of vegetative propagation. More than 4,000 years ago, it was tried in China and, because of its continued use in that country, it has traditionally been called Chinese layering. It is also sometimes referred to as marcottage from the great era of French gardening in the late seventeenth and eighteenth centuries.

An unpruned stem on a normally growing woody plant is induced to develop roots by restricting the stem about 6–9 in behind its growing tip and then excluding light. This combination will cause the initiation of roots, and, if the roots are surrounded by moist, warm soil, then they will develop. The stem is then severed from the parent and established as a new plant.

As a technique, air layering can be used on a wide range of plants that have woody stems, and it is a useful way of producing plants without specialized equipment or disturbing the parent plant. It is carried out either in spring on the matured wood of the previous season, in which case layering is close behind the growing point, or in the late summer on the hardening shoots of that season's growth.

Select a stem of the current year's growth. Trim any side-shoots off the stem for about 6–12 in behind the tip.

Then girdle the stem (see page 107) so that food and hormones build up in the region where rooting is required – usually abut 9 in behind the tip. Treat the stem or the cut surfaces with hormone powder to improve rooting.

The most effective rooting medium is sphagnum moss as it holds water, is well aerated and is readily manipulated. Soak the moss overnight so that it is completely moist. Take two large handfuls of moss and squeeze it, then work it into a ball so that all the fibres are interwoven. When it reaches about 2½ in in diameter, split it in half, using the thumbs, in the same way as an orange is divided. Place the two halves around the treated area of the stem and knead them together again so that the moss stays firm.

Hold the moss in place with a square of black polythene, and secure it to the stem with sticky insulating tape. It is important to ensure that moisture cannot run down into the moss and so waterlog it. So, turn the tape spirally round the ends of the polythene, overlapping it until it covers part of the stem.

The black polythene will retain moisture, maintain a warm environment, exclude light and allow gases to permeate the moss.

The layered stem will usually take at least a growing season to produce adequate roots.

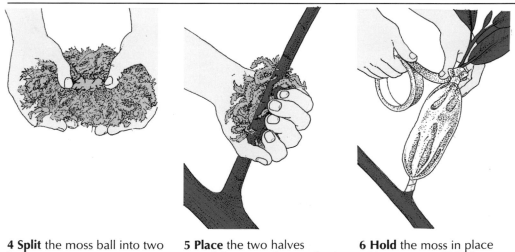

4 Split the moss ball into two in the same way as an orange is divided.

5 Place the two halves round the treated area of the stem. Knead them together once more.

6 Hold the moss in place with a square of black polythene. Secure with stick insulating tape.

Prune back any new growth on the rooted layer towards the end of the dormant season. Then cut off the stem with secateurs just below the point of layering and remove the black polythene square.

The most critical stage when air layering is to establish the rooted layer successfully. Fill a pot with John Innes No. 1 compost or similar (see page 23). Slightly loosen the moss ball and roots. Place the rooted layer in the pot. Firm so the roots are in contact with the compost. Do not firm too heavily and so compress the moss ball and roots. Label and place in a protected environment until further root growth develops and the new plant is established.

1 Trim any leaves and side-shoots for about 6–12 in behind the tip of the stem to be layered.

2 Girdle the stem to encourage root formation (see page 107). Then treat it with rooting hormone.

3 Squeeze two handfuls of wet sphagnum moss together. Knead into a ball of about 2½ in diameter.

7 Prune back any new growth on the rooted layer towards the end of the dormant season.

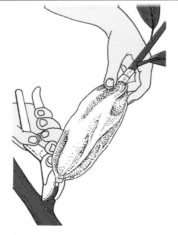

8 Cut the stem just below the point of layering, using a pair of secateurs. Remove the black polythene square.

9 Loosen the moss ball and roots slightly. Then pot in John Innes No. 1 compost. Firm in gently and label.

Tip layering

Tip layering is a specialized technique used for the various members of the genus *Rubus*, such as blackberries and loganberries. If the growing tip of such a plant is buried in the soil, it will naturally swell, develop roots and establish itself. This phenomenon is modified to suit the gardener.

Tip layering is an invaluable technique for propagating a few plants, as it can be carried out in the garden on one part of a plant without disturbing the flowering or fruiting ability of the rest of it.

The members of the genus *Rubus*, especially the fruiting varieties, are prone to virus infections, and they should be propagated only from known virus-free stocks.

Select a new strong stem as it develops from the crown of the plant during the spring. As soon as it reaches 15–18 in long, pinch out the tip to encourage branching. The growth of these stems is rapid and vigorous, and soon the stems can be pinched out again. Continue to do this until midsummer, when about six to eight tips should be well developed.

It is at this stage that the stems can be layered. However, as the roots of these particular plants are fine, fibrous and easily damaged, the soil for layering should be well prepared so that once the layers have rooted they can be lifted with minimum damage to their root system.

1 Pinch out the tip of a 15–18 in basal shoot of the current year's growth.

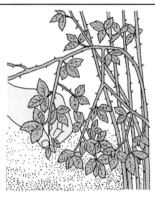

2 Continue to pinch out the tips until six to eight tips have developed.

3 Cultivate the soil well. Add peat and grit to the top 6 in of the soil.

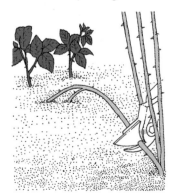

7 Cut back the original stem at the crown of the parent plant in September.

8 Cut off the rest of the stem once the rooted layer has dropped its leaves.

9 Shorten some of the top growth that the layer may have made.

**Some plants responding to
tip layering**
Blackberry
Boysenberry
Loganberry

Rubus phoenicolasius
(Japanese wineberry)
R. linkianus 'Plenus'
R. ulmifolius 'Bellidiflorus'
Veitch berry

Dig and cultivate the soil thoroughly — if possible dig in a deep layer of organic matter to conserve moisture and prevent the roots drying. Improve the top 6 in by adding some peat and grit.

Pull down a stem and make a mark where its tip touches the ground. Start digging a trench at this spot, making it 4 in deep. Give it smoothed vertical sides except for one which should be sloped towards the parent plant. Smoothed vertical sides will help any shoots to grow vertically and so produce a manageable plant.

Place the tip in the deepest part of the trench and pin it down with a heavy wire staple. Replace the soil, firm and water.

In about three weeks, shoots should appear above ground level.

In September cut back the original stem at the crown of the parent plant, so that the layer can establish as an independent plant.

Cut off the rest of the original stem and shorten the growing tip before lifting the rooted layer in autumn and after leaf-fall. Replant the layer immediately and label it.

Protect any layers that cannot be replanted straight away by wrapping their roots in damp newspaper, which is then placed in a plastic bag. Close the bag and tie the neck tightly so the roots will not dry out.

Propagation by tip-layering in this way can be repeated each year.

4 Pull down a stem. Dig a trench 4 in deep where its tip touches the ground.

5 Place the tip in the deepest part of the trench. Pin it down with a staple.

6 Replace the soil, firm and water, using a can with a coarse rose.

10 Lift the layer very carefully to avoid damaging its fine fibrous roots.

11 Plant the rooted layer at once in well-cultivated soil. Label it clearly.

12 Place layers that cannot be planted at once in damp newspaper in a plastic bag.

Stooling

Stooling is an entirely artificial system of propagating plants by layering, because a plant is grown just to develop new plants, which it will do year after year.

It is a technique that is principally used to produce specialized rootstocks that control the vigour and size of a tree, especially fruit trees. However it can be employed on any plant that will respond to severe annual pruning.

Plant a rooted layer, cutting or seedling in well-cultivated ground and label it. Establish for one growing season. Never use a plant that has been grafted as it will reproduce the rootstock and not the grafted variety. All clonal fruit-tree rootstocks should be designated EMLA, which indicates they are free from virus disease. During the early part of the dormant season, cut back the rootstock, leaving about 1–2 in of the stem above ground level.

In the following spring, earth up any shoots as soon as they are about 6 in long, so that the whole plant is covered. Work the soil down between the shoots so that each shoot is completely surrounded by soil. Do not delay earthing up as one of the most critical factors in stooling is excluding light from the plant at an early stage.

As the shoots grow, continue to earth them up until each shoot is buried to about 9 in.

No further action will now be necessary, unless the summer is particularly dry, in which case the soil should be watered to encourage the roots to develop. Just keep the soil around the stems warm and moist. Too much water will depress temperatures.

In early winter, after the leaves have fallen and the shoots are fully dormant, gently fork away the soil back down to the original level, so exposing the stool and its shoots. The base of each shoot should have produced roots.

Remove these rooted layers from the parent stool. With a pair of secateurs cut them flush with the stool, so that no stub remains. Replant them immediately and label them clearly.

After removing and replanting the rooted layers, cultivate the soil around the stool and clean off any residual earth on the stool itself so that it is fully exposed to the elements. This is necessary to ensure that the buds receive adequate winter chilling and will therefore break evenly the following spring, when the whole process is repeated, except that a general fertilizer is added at abut 4 oz to the square yard. Successful production will depend on looking after the stool.

4 Continue to earth up until each shoot is buried to about 9 in in soil.

5 Fork away all the soil in early winter once the leaves have fallen.

6 Cut off any rooted layers flush with the stool, using a pair of secateurs.

Some plants produced by stooling
Apple – East Malling series;
   Malling Merton series
Chaenomeles

Lilac
Plum – Brompton;
   Myrobalan B; St Julien A
Plum, coloured-leaf
Quinces 'A' and 'C'

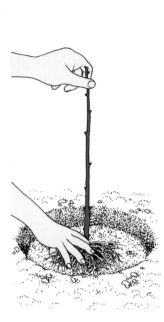

1 Plant a labelled rooted layer in well-cultivated soil. Allow it to establish for one growing season.

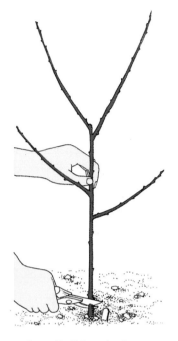

2 Cut off all but the bottom 1–2 in of the stem during the early part of the dormant season.

3 Cover the plant with earth as soon as the shoots are 6 in long. Firm the soil between each shoot.

7 Replant each rooted layer immediately into the open ground and label it.

8 Cultivate the soil around the stool, which should then be cleaned well.

9 Add a general fertilizer, at about 4 oz to the square yard the following spring.

French layering

French layering is an extension of stooling and, as such, it also requires a specially grown plant from which new plants can develop. Like stooling, it produces an annual crop of rooted layers, but initially it takes longer for the sequence to become established.

Cultivate some ground by digging deeply and adding some organic matter, peat and grit. Plant a rooted layer and label it. Allow it to establish for a growing season. During the early part of the dormant season, cut it back to 1–2 in above ground level.

Leave the plant to grow undisturbed during the following growing season.

After the leaves have dropped, reduce any shoots to a manageable number – say, at the most, eight strong shoots. Cut back their tips so that each stem is the same length; then peg them down horizontally over the ground. By positioning the stems horizontally early in the winter, the buds will break evenly along the entire length of the shoots in spring.

In spring, unpeg the stems when their shoots are about 2–3 in long. Cultivate the

1 Plant a rooted layer and label it. Allow it to establish for a growing season.

2 Cut off all but 1–2 in of the stem in the early part of the dormant season.

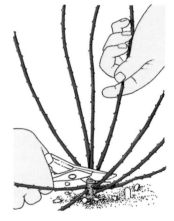

3 Reduce the shoots to eight after leaf-fall the following year.

7 Drop each stem into a trench 2 in deep. Cover, leaving shoot tips exposed.

8 Earth up the new shoots as they grow until the mound is about 6 in high.

9 Fork away the soil mounds at the end of the growing season.

**Some plants that can be propagated
successfully by French layering**
Cotinus coggygria (Rhus cotinus) –
  varieties
Dogwood, coloured-bark
Maple – some species

ground. Space out the stems evenly in a star arrangement around the stool, and make a trench 2 in deep under each stem. Drop each stem into its trench and cover it until only the tips of the new shoots are exposed.

Earth up the new shoots as they grow, always keeping the tips exposed, until the mound is about 6 in above ground level. Water the layers only if the weather is particularly dry. Do not overwater.

The central stool meanwhile will be producing a fresh crop of shoots, and it is these that will provide the next year's stems for pegging down.

Gently fork away the soil mounds, after leaf-fall. Cut away the rooted stems flush with the stool, avoiding the current year's growth. Divide each stem into individual plants with their own root systems. Plant these out immediately; label and water in.

Reduce the current year's growth on the stool to, say, eight strong shoots, and prune to an equal length. Peg these down on the ground so that the sequence can continue.

4 Cut back the growing tips so each shoot is the same length.

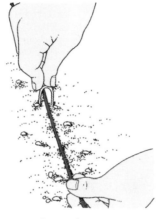

5 Peg down these stems horizontally over the ground.

6 Unpeg the stems when their shoots are 2–3 in long. Cultivate the ground.

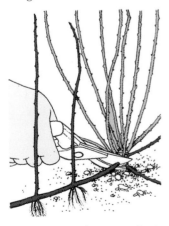

10 Cut away the stems flush with the stool. Ignore any new growth on the stool.

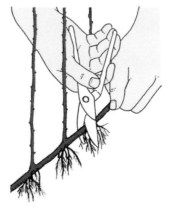

11 Divide each stem into individual plants with their own root systems.

12 Replant immediately in the open ground. Label and water in well.

Dropping

This technique is used to propagate numerous heaths and heathers, dwarf rhododendrons and other related plants, and any shrub of suitable habit that is not readily propagated by other methods.

A plant that is already mature, and possibly even straggly, is dug up with a reasonably sized root ball. It is then completely buried so only the tips of the branches show. These branches will root, and they are then lifted, separated, and planted out to establish as new plants.

Dropping is not necessarily the most desirable method of propagating plants as often the resulting layers are less shapely than those produced by cuttings, but it is easy and simple to follow.

Prepare a plant in the dormant season by pruning it rigorously to encourage new, strong-growing stems with a high capacity to produce roots. Older, non-pruned stems will respond less satisfactorily to propagation but will nevertheless usually regenerate.

Dropping is normally done in spring before growth begins, but once the ground is no longer frozen, so the soil can be broken down to a tilth.

Cultivate some soil and incorporate peat and grit if the soil is heavy and likely to become waterlogged. This will lighten it and improve aeration.

Excavate a hole large enough for the whole plant to be "dropped" into it, leaving only the tips of the stems visible. Dig the base of the hole well to allow efficient drainage, otherwise waterlogging will discourage successful rooting.

Lift the plant with as complete a root ball as possible. Place it in the hole and arrange its branches in any of three different patterns within the hole; the choice will depend upon the habit of the plant under consideration (see opposite).

Whichever pattern of dropping is used, ensure that only 1 in or so of the stem tips are exposed: greater exposure of stem will cause the new growth to be leggy. Cover the plant with soil, firm in and label. Water during the growing season if the soil dries.

Lift the whole plant in autumn. The branches will have rooted, usually fairly close to the soil surface. Cut away each division and plant it out or pot it up. Label clearly. Discard the old stool.

1 Prune a plant in the dormant season to induce stems with a high capacity to produce roots.

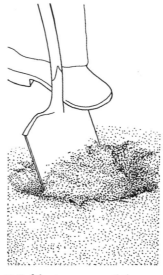

2 Cultivate some soil the following spring. Add peat and grit if necessary. Dig a very large hole.

3 Lift the plant with as complete a root ball as possible. Place it in the hole.

Some plants that can be propagated by dropping
Andromeda
Calluna
Cassiope

Daboecia
Erica
Hebe sp. – whipcord types
Kalmia angustifolia
Ledum

Pernettya
Phyllodoce
Rhododendron – dwarf and small-
    leaved types
Vaccinium

WAYS TO BURY A PLANT

Compress the branches into a single row towards the middle of the plant if the branching is sparse. This pattern saves space and makes weeding easy. It should not be done if the plant produces a thick mass of stems as rooting will not be satisfactory if there is insufficient room for the roots to develop.

In those plants with brittle branches, work the soil down among them so that each stem is surrounded. This pattern is more difficult to keep weed-free.

The traditional pattern is to excavate a bowl-shaped hole and to push out all the branches to the perimeter, and fill in the middle of the hole with soil. This pattern is easy to weed but is wasteful of space.

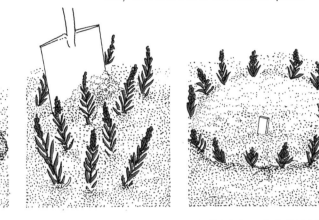

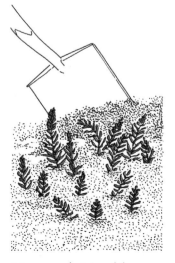

4 Leave only 1 in of the stem tips exposed when covering the plant with soil. Firm it in and label.

5 Water during the growing season if the soil dries.

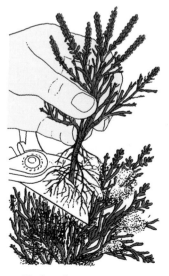

6 Lift the plant in autumn. Cut away each rooted stem and plant it out or pot it up. Label clearly.

Stem cuttings

Growing plants from stem cuttings is by far the most popular method of vegetative propagation. However, only in the last 150 years, since the availability of cheap materials and the development of greenhouses, cold frames, etc., has it played a significant part in vegetative propagation.

The main difficulty about taking stem cuttings is that a stem, separated from its parent plant, has to survive while it initiates and develops roots and establishes itself as a new plant. This distinguishes the technique from layering (see page 106), in which the stem is not separated from the parent plant until after a new root system is produced.

Because a stem cutting has no support system from the parent plant, it is necessary for the gardener to supply this himself. He should ensure that the propagating environment will not only enhance root development but also maintain the cutting until it is self-supporting.

Selecting a suitable stem

The ability of a stem to produce roots will be dependent on the age of the parent plant and its particular variety, and on the stem to be propagated (see pages 104–5). The gardener should try to propagate from a plant that has recently matured and is a relatively new variety rather than an old plant or old variety of plant.

The parent plant should be pruned rigorously to encourage it to produce fast-growing vegetative shoots from which stem cuttings will be made, as these shoots are most likely to produce roots. The harder the plant is pruned, the faster will be the new growth. The whole success of propagation by stem cuttings depends on the ability of the stem to produce roots – and if this is absent or at a very low level, then the stem should be discarded.

A stem's ability to produce roots may have seasonal fluctuations, but this depends on the condition of the stem, that is whether it is soft wood, hard wood, etc. A softwood cutting is taken soon after the buds have started growing in spring, and it has a greater ability to produce roots than a hardwood cutting, which is taken at the end of the growing season. However, because the softwood stem is still immature, it is more susceptible to water loss, rot and disease, and it therefore requires a highly controlled environment in which to develop.

A stem cutting's food reserve is used not only to initiate roots but also to maintain the cutting until it is fully established as a new plant. The size of the reserve depends on the condition of the stem: a cutting from a mature (hardwood) stem will be able to survive much longer than an immature (softwood) cutting. A cutting should therefore be encouraged to develop roots as quickly as possible to avoid exhausting its food reserve. It should also be exposed as little as possible to variable weather conditions to prevent it drying out – leafy cuttings being particularly prone to water loss.

A cutting should be taken from a fast-growing stem from the current year's growth at the correct season for the condition of the stem (for example, green wood in early summer, hard wood during the dormant season). It should then produce roots quite readily, without the artificial aid of rooting hormones; if it is dipped in hormone, the treatment is likely to have little or no effect. However for plants that are difficult to root, dip stem cuttings into a rooting hormone, or wound them, to stimulate root production.

Environmental control

The rate at which a stem cutting develops its roots is dependent on the temperature around it. The processes controlling root initiation are essentially chemical; the higher the temperature the faster the chemical reaction and thus root production. However, if a whole cutting is kept warm, its tip will grow and food will be diverted from the important function of forming roots. Its food reserves may then be used up before the cutting has become self-supporting. Therefore, a cutting requires two temperatures: a cool, aerial environment to keep tip growth to a minimum, and warmth below to encourage root production.

The exact temperatures vary with the condition of the stem and how susceptible it is to water loss. Softwood cuttings require bottom heat of about 21°C/70°F and as cool an aerial temperature as practical – a mist unit is ideal. Hardwood cuttings, on the other hand, are propagated outdoors where the soil is quite warm enough and the air, even when frosty, is not too cool. Greenwood, semi-ripe and evergreen cuttings need a warm, humid environment. This can be supplied by placing a small pot filled with water inside the pot that has the cuttings and compost in it and putting them under a polythene tent, or by placing the pot that has the cuttings and compost in it inside a larger pot filled with moist peat and putting them under a polythene tent. However, the main disadvantage of these two environments is that it is easy for the gardener to overwater and so kill the cuttings through rotting and fungal diseases unless extreme care is taken. Cold frames, closed cases and polythene tunnels are therefore preferable.

Different stem conditions
A stem cutting can be divided into five wood conditions for the purposes of propagation.

Softwood cuttings have leaves and are made from the first flush of growth in spring. Their stems are normally very soft because they have grown extremely rapidly; and they require sophisticated environmental controls to minimize water loss and so ensure their survival until they become established.

Greenwood cuttings are made from the tips of the leafy stems during early to mid-summer. Their stems are soft, although harder than softwood cuttings, and they should be propagated in a controlled environment, such as a closed case.

Semi-ripe cuttings are made in late summer from stem growth that has slowed and hardened but is still actively growing. Although these leafy stems are subject to water loss, they can survive under less rigorous environmental controls than softer wood cuttings.

Ripewood cuttings are stems taken from evergreen plants during winter. They have almost hardwood stems but, because the are leafy, they are not entirely dormant and will require some degree of environmental control.

Hardwood cuttings are made from leafless dormant stems of deciduous plants. They require only minimal environmental control for survival.

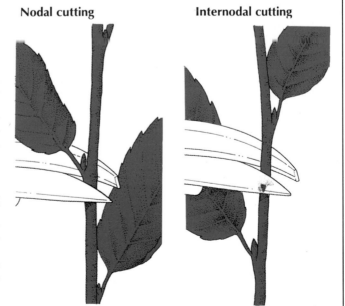

WHERE TO CUT A STEM
A nodal cutting has its basal cut just 1/8 in below a bud, or node. It is the traditional place to cut soft, immature stems as a stem just below a node is harder and more resistant to fungal rots than a stem cut further away from the nodes. This latter cut is known as an internodal cut, and it is used mainly for more mature, woodier stems.

Other methods of taking a cutting are leaf-bud, heel or mallet cuttings (see pages 122–5) or vine-eye cuttings (see pages 144–5). The index at the back of the book will refer to these methods when appropriate.

Nodal cutting

Internodal cutting

Making a stem cutting 1

Leaf-bud cutting

Leaf-bud cuttings can be taken from any type of stem – soft wood, green wood, semi-ripe wood, hard wood or evergreen. Each cutting consists of a leaf, a bud in its leaf axil and a very short piece of stem. The leaf supplies food to support the cutting and the regenerative processes; the bud provides the basis for the new stem system; and the piece of stem is where the first roots are produced.

To be successful, the gardener must use stems that have a high capacity to produce roots. Therefore, prune the parent plant rigorously should it be a woody plant. This will encourage new stems, which will grow rapidly and so have a high rooting potential.

For leaf-bud cuttings, select one of these new stems with an undamaged leaf that is fully expanded and mature. If the leaf is immature, the cutting will complete leaf growth before it starts producing roots, and this increases the chances of peripheral problems, such as rotting. Also, ensure that there is a viable bud in the leaf axil. (For example, some virginia creepers do not have a bud in every leaf axil.)

Make the cuttings with a razor blade, knife or secateurs, depending on the hardness of the stem. Cut close above the bud so that as small a snag as possible is left. This minimizes the likelihood of rotting and die-back, which might endanger the bud.

Make the basal cut about 1–1½ in below the top cut so that sufficient stem is available to anchor the cutting firmly in the cuttings compost. This is especially important with plants that have big leaves and are liable to rot.

Plants with big leaves are also difficult to plant at a realistic spacing, so reduce their leaf area either by removing some of the leaf or by rolling the leaf and placing a rubber band round it so that it takes up less room. Dip the cutting in a rooting hormone.

In a pot filled with cuttings compost make a hole with a dibber. Plant the cutting with its bud about level with the compost surface. Firm sufficiently to prevent rocking. Label and water in. Place hardy cuttings in a cold frame and less hardy cuttings in a well-lit, more protected environment, such as a mist unit or closed case.

Leaf-bud cuttings

1 Prune the parent plant, if suitable, to encourage new stems with a high rooting potential.

2 Select a new stem with an undamaged mature leaf and a viable bud in its axil, later in the season.

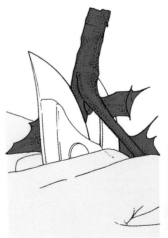

3 Make an angled cut just above the bud.

Some plants suitable for propagation by leaf-bud cuttings
Aphelandra
Camellia
Clematis

Ficus elastica (rubber tree)
Haberlea
Ivy
Mahonia
Ramonda
Vine

DOUBLE LEAF BUDS

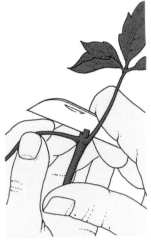

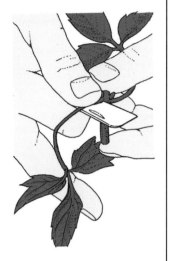

1 Cut a stem with leaves that grow opposite each other just above a bud and again 1–1½ in below it.

2 Split the stem down the middle, using a sharp knife, to make two cuttings.

OR Remove one leaf. Dip the cutting in a rooting hormone. Then plant and label clearly.

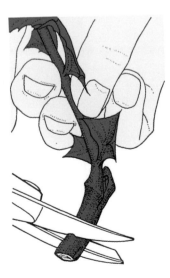

4 Make a straight cut 1–1½ in below the top cut.

5 Roll up or cut a large leaf. Dip the leaf-bud cutting in a rooting hormone.

6 Plant in a pot with its bud about level with the compost surface. Firm well. Water in.

123

Making a stem cutting 2

Heel cuttings

The taking of heel cuttings is a traditional way to propagate plants from stem cuttings. It is a widely used method of removing a stem cutting from a plant, and it is quite possible to make heel cuttings of softwood, greenwood, semi-ripe, hardwood or evergreen stems.

A young side-shoot is stripped away from its parent stem so that a heel, that is a thin sliver of bark and wood from the old stem, also comes away at the base of the cutting.

The reason for taking a stem cutting with a heel is to give the cutting a firm base so that it is well protected against possible rot. It also exposes the swollen base of the current season's growth, which has a very high capacity to produce roots.

Heel cuttings are often used for stem cuttings that take some time to develop roots, for example those that are planted in autumn and have to survive through the winter before rooting, or those hardwood cuttings that are planted in a cold frame. Heel cuttings are also made from softwood and greenwood stems that are left to develop in partially controlled environments, such as a propagator. Heel cuttings can be taken at any time of the year.

Hold the bottom of the side-shoot between the thumb and forefinger and pull down sharply so it comes away with a long tail.

If the side-shoot does not pull off readily, place a knife blade in the angle close against the parent stem and cut away the side-shoot with a heel.

Trim the tail on the heel and any leaves near it. Remove some of the tip on semi-ripe and hardwood cuttings. Dip the cutting in a rooting hormone.

Heel cuttings

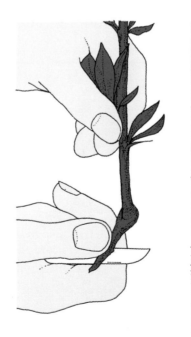

1 Hold the bottom of a side-shoot between the thumb and forefinger. Pull down sharply.

2 Neaten the long tail on the heel and any leaves near it. Dip the basal cut in a rooting hormone.

3 Make a hole in the soil or compost. Plant the cutting. Label it and water with a fungicide.

Plant hardwood cuttings straight into the ground; semi-ripe and sub-shrub cuttings in a cold frame; and less hardy cuttings in the protected environment of a propagator. Label, and water in with a fungicide.

Mallet cuttings

Mallet cuttings have a hardwood plug at the base of each cutting to guard against rotting organisms. Their use is restricted to semi-ripe and hardwood cuttings, and they are especially successful for many *Berberis* when propagated in the autumn and planted in cold frames.

Mallet cuttings are most successfully made from stems with a feathered habit, that is from a stem with small side shoots.

Prune back the parent plant in winter to encourage vigorous stem growth, which has a high capacity to produce roots.

Take mallet cuttings from these new stems in the later part of the growing season. Cut horizontally with scissor-type secateurs across the parent stem immediately above a suitable side-shoot. It is important to make this top cut as close to the side-shoot as possible because the longer the snag the greater the likelihood of die-back and hence potential rotting.

Make a further horizontal cut about ¾ in below the top cut so that the side-shoot is isolated with a small "mallet" of parent stem. Split this piece of mallet with a knife if it is thick. Trim any leaves at the bottom.

Dip the base of the mallet stem in a rooting hormone. Make a hole large enough to take all the mallet and part of the side-shoot with a dibber. Plant semi-ripe cuttings in a cold frame and hardwood cuttings in the open ground. Label, and water with a fungicide.

Mallet cuttings

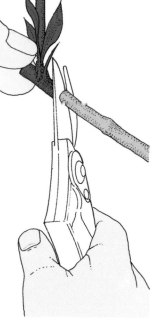

1 Cut horizontally with secateurs across the parent stem just above a side-shoot.

2 Make a basal cut about ¾ in below the first cut. Remove any leaves at the bottom.

3 Dip the basal cut in a rooting hormone. Plant the cutting and label it. Water with a fungicide.

125

Soft woods 1

Soft wood is the most immature part of a stem, and, when propagating, it is the most difficult kind of cutting to keep alive. However, soft wood does have the highest capacity of all kinds of stems to produce roots: the younger and the more immature the cutting, the greater will be its ability to develop roots, and so propagate successfully.

Soft stem growth is produced continuously at the tip of any stem during the growing season. As it matures, the stem gradually hardens and becomes woody. The faster the growth at the tip, the more stem without wood will be present.

Softwood cuttings, then, are taken in spring from the fast-growing tips of plants. Growth when the buds first break is remarkably rapid, and, if cuttings are made as soon as there is sufficient growth, the highest rooting potential will be available. Tips of plants taken slightly later in the season, around the beginning of June, will be slower growing, more mature and have a lower capacity to root, and they are referred to as greenwood cuttings.

It is possible to obtain softwood cuttings later in the season by forcing the plant, that is by increasing temperatures well above the norm, which will accelerate growth. This can be done, for example, by placing deciduous outdoor plants in the warmth of a greenhouse, or by placing house plants in a temperature of about 29°C/85°F. The very highest capacity to produce roots can be achieved by pruning the parent plant vigorously in winter, which will encourage rapid

1 Prune woody plants hard in winter to promote stems with a high capacity to produce roots.

2 Fill a container with cuttings compost. Firm to within 3/8 in of the rim.

3 Cut the fast-growing tip off a stem in the early morning in spring.

growth; by increasing the temperature around the plant in spring; and by then taking the cutting as soon as sufficient tip growth is available.

Softwood cuttings are extremely susceptible to water loss. Their immature leaves have not fully expanded and so have not completely developed their own mechanisms for reducing water loss. Even a relatively minor water loss will hinder the roots developing. By the time a cutting is wilting all root development will have ceased.

The secret of success is to collect the cuttings in small batches and to maintain them in a fully turgid condition before planting them.

Fill a container with cuttings compost and firm to within 3/8 in of the rim.

Take a cutting in early morning when the stem is fully turgid. By mid-afternoon normal water loss from the plant will exceed uptake, and the plant will be under water stress.

If the stem has grown less than 4 in since bud-break, remove it with a heel, that is with the swollen portion at the base of the stem that had the very fastest growth when the new stem started to develop at bud-break.

Place the cutting in a polythene bag or a bucket of water immediately. Keep the polythene bag shaded to avoid "cooking" the cutting – a major cause of failure.

Plant the cutting as soon as possible. If it cannot be dealt with quickly, keep it cool in the salad-box area of a domestic refrigerator, where the low temperature will prevent excessive water loss.

4 Place the cutting at once in the shade in a polythene bag or a bucket of water.

5 Cut the base of the stem 1/8 in below a leaf joint if the cutting is more than 4 in long.

OR Trim the tail if the cutting was taken with a heel.

Soft woods 2

If there is more than 4 in between the tip and the base of the cutting, place the cutting on a sheet of glass and make a nodal cut – that is a cut at the base of the stem just (⅛ in) below a bud or leaf joint. This provides as hard and solid a surface as possible and will help prevent rotting.

If the cutting was taken with a heel, neaten the tail that will also have come away with the stem.

Remove the leaves on the bottom third of the cutting, which will be reliant on the remaining leaves to produce sufficient food to keep it alive until its roots are fully established.

Dip the base of the cutting in a proprietary hormone rooting powder containing a fungicide to protect it against rotting, although soft-wood cuttings do not need to be treated with a rooting hormone.

Make a hole with a dibber and plant the cutting up to its leaves in the compost, taking care not to damage the base of the cutting. Plant cuttings so their leaves do not touch. Label and firm them in by watering from above the compost, using a watering can with a fine rose; pressing by hand may damage the cuttings.

Place the cuttings as quickly as possible in a well-lit propagating environment such as a mist unit, a closed case or a polythene tent that will conserve moisture within the cuttings.

The advantage of a mist unit is that it keeps the top of the cuttings cool, whereas air

6 Remove the leaves off the bottom third of the cutting. Dip the basal cut in a powder fungicide.

7 Make a hole with a dibber in the compost. Insert the cutting up to its leaves. Plant any more cuttings.

8 Label the cuttings clearly. Water from above the compost. Place in a well-lit protected area.

temperatures are high in a closed case or polythene tent. A high aerial temperature will force the cuttings to grow upwards, and food will be diverted from the important function of root initiation. If the cuttings are shaded to reduce the aerial temperature, then the light intensity penetrating to the leaves of the cuttings is decreased, and this reduces food production and hence the rate of regeneration. The problem then becomes a vicious circle that is not easy to resolve without a mist unit, which maintains both a cool, aerial environment and high water status within the cuttings.

Softwood cuttings, because they are the immature part of a plant, are susceptible to all the vagaries of their environment: so the longer they take to root, the greater are the chances of them succumbing to some outside influence. Thus the speed with which they regenerate is vital. The rate of production will be dependent on the temperature surrounding the base of the cuttings; in general, the higher the temperature, the faster the roots are produced. Best rooting will occur with a temperature around the base of the cuttings of 21–24°C/70–75°F.

Harden off the cuttings once they have rooted successfully, gradually weaning them from their controlled environment; finally pot them up in John Innes No. 1 compost or similar (see page 23) and label.

9 Harden off the cuttings gradually when they have rooted.

10 Pot them in John Innes No. 1 compost once they are weaned and label.

Green woods

The essential but subtle distinction between softwood cuttings and greenwood cuttings is their speed of growth. Externally they may appear to be very similar, but greenwood cuttings are taken from the soft tip of the stem after the spring flush of growth has slowed down. The stem is, then, slightly harder and woodier than for softwood cuttings because the tip is not growing away from the hardening stem so fast as it was in the spring. In fact, as the season progresses greenwood cuttings become harder and harder. However, it should be emphasized that greenwood cuttings require just as much environmental control as softwood cuttings.

Use greenwood cuttings to propagate a wide range of trees and shrubs, such as gooseberries, that will root easily, and the majority of herbaceous plants, such as chrysanthemums. However, plants that are difficult to root should be propagated from softwood cuttings and not from greenwood cuttings, which have a slightly reduced ability to develop roots.

Prune back woody plants rigorously in winter to encourage rapidly grown stems with a high capacity to root, which can be used for propagation in the growing season.

Take cuttings from these stems once their growth rate has begun to decline, which will usually be about the beginning of June for most outdoor plants.

Fill a container with cuttings compost and firm to within 3/8 in of the rim.

Take a cutting from a fully turgid stem with all its current season's growth, early in the morning. Place the cutting immediately in a bucket of water or a polythene bag in a shaded position, because it is vitally important to maintain the turgidity of a greenwood cutting. Water loss will hinder rooting.

Place the cutting on a pane of glass and with a knife trim the cutting to about 3–4 in. The cutting's length will very much depend on the amount of soft growth available. Discard the leaves from the bottom half of the cutting. Dip the base of the cutting in a rooting hormone powder of "softwood" strength containing a fungicide.

Make a hole in the compost and plant the cutting up to the leaves. Label it and water. Place in a closed case, mist unit or polythene tent, to prevent excessive water loss and wilting. Ensure the cutting receives adequate light

4 Place the cutting on a pane of glass. Reduce to 3–4 in, using a sharp knife.

5 Trim the leaves from the bottom half of the cutting, cutting flush with the stem.

6 Dip the base of the cutting in rooting hormone containing a fungicide.

Some plants suitable for propagation by greenwood cuttings
Berry fruits
Ceanothus

Chrysanthemum
Delphinium
Forsythia
Geranium (*Pelargonium*)
Gooseberry

Philadelphus
Vines – fruiting and ornamental

to make the food needed for root production, as the cutting has little or no food reserve.

Rooting should take between three and eight weeks. Take the rooted cutting out of the propagating environment but keep it well protected, gradually hardening it off. Then re-pot in John Innes No. 1 compost or similar (see page 23), and label clearly.

1 Prune woody plants in winter to encourage strong, vigorous stems with a high capacity to root.

2 Fill a container with cuttings compost. Cut off all of a vigorous stem early in the morning.

3 Place the cutting at once in a bucket of water in the shade.

7 Make a hole with a dibber in the compost. Insert the cutting up to its leaves. Label the container.

8 Water. Place in an environment that will control water loss and give shaded light.

9 Harden off the rooted cutting. Repot and label as soon as it has established.

Semi-ripe woods

During the late summer, annual stem growth slows down and plant stems become harder. Cuttings taken at this time are called semi-ripe cuttings. As they are thicker and harder than softwood cuttings, they are more capable of survival. However they are still susceptible to the same problems of water loss because the cuttings carry leaves.

Semi-ripe cuttings have relatively high levels of stored food and can therefore survive and produce roots in poor light.

Many deciduous plants, such as deutzias, that root fairly easily are propagated from semi-ripe cuttings, and some evergreen plants can also be increased in this way.

Prune the parent plant at the start of the dormant season, so that strong, fast-grown stems are available for propagation the following season. These will have a greater ability to produce roots than unpruned stems.

Prepare the soil in the cold frame by digging deeply. Add peat and grit to improve drainage and its water-holding capacity. Cover the surface with a 1 in layer of fine (builder's) sand to make a better rooting medium. If only a few cuttings are to be taken, fill a container with cuttings compost and cover with a 1 in layer of fine sand.

Take semi-ripe cuttings from a main stem with all the current season's growth or from side-shoots if the main stem has feathered growth, that is a series of small side-shoots growing on the main stem.

Remove the tip of the stem if it is soft, but leave it if the apical bud has set and growth has ceased for the year. Shorten the cutting with secateurs to 4–6 in long, depending on the vigour of the particular plant.

Cut off the lower leaves flush with the stem so that about 2 in of stem is clear at the base.

Treat the basal cut surface with a rooting hormone powder containing a fungicide. The strength for semi-ripe cuttings is 0.4 per cent IBA.

Make a hole with a dibber in the soil in a cold frame. Plant the cutting about 1½ in deep, so its base just enters the soil below the sand layer. Space the cuttings as close as is feasible, but this is not likely to be much less than 3–4 in apart. Label them clearly.

Water in the cuttings; this also firms the sand around the cutting. Close the frame tightly and shade it to prevent the leaves from scorching. Air the cold frame from the lower end if the temperature rises above 27°C/80°F. Water sufficiently to rewet and develop high humidity should the prevailing conditions become dry.

As semi-ripe cuttings are usually deciduous, they will drop their leaves in autumn. At this stage remove all the fallen leaves from the frame so that they do not rot.

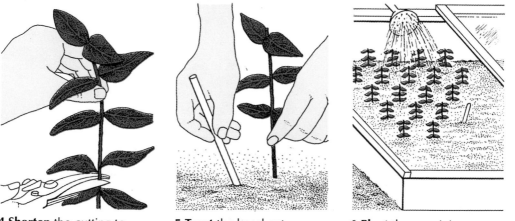

4 Shorten the cutting to 4–6 in. Cut the leaves off the bottom 2 in of the stem.

5 Treat the basal cut surface with some rooting hormone. Dibble in a cutting 1½ in deep.

6 Plant the remaining cuttings 3–4 in apart. Label and water. Seal and shade the frame.

Some plants that can be propagated from semi-ripe cuttings
Currant, flowering
Deutzia
Diervilla
Dogwood, coloured-bark
Forsythia
Philadelphus
Plum, coloured-leaf
Weigela

Insulate the frame with matting once the leaves have fallen to protect the cuttings against frost in winter. Once the cuttings have no leaves, light is not necessary for the plants to manufacture food and so the matting can be left in place all the time.

Rooting may start fairly quickly if the weather is mild; otherwise it will occur during the late winter or spring.

Leave the rooted cuttings *in situ* during the following growing season. Feed them regularly with a liquid fertilizer and water when dry. Remove insulation and air the cold frame by raising its lid during the day as soon as the danger of frost is over. Eventually remove it altogether. Lift the new plants and transplant them after leaf-fall in the autumn. Label the new plants clearly.

1 Prune the parent plant at the start of the dormant season to encourage strong stems to grow.

2 Prepare the soil in the cold frame by digging deeply. Add peat and grit. Cover with 1 in fine sand.

3 Cut off a shoot with all its current season's growth in later summer. Remove the tip only if it is soft.

7 Rewet the soil if it dries. Remove any fallen leaves. Insulate the cold frame with matting.

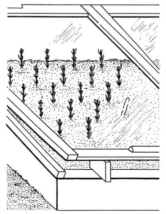

8 Remove the insulation and air the frame once the danger of frost is over. Apply a routine liquid feed.

9 Lift the new plants and transplant them once their leaves have dropped in autumn. Label them clearly.

133

Evergreens 1

The propagation of evergreen trees and shrubs from stem cuttings is a traditional and valuable method. Evergreen cuttings are taken from stems of very ripe wood, that is, almost hard wood. They cannot be regarded as hardwood cuttings as they are not leafless and are not fully dormant because of their evergreen habit. Because they have leaves, they need sufficient environmental control to prevent excessive water loss.

In winter, prune the parent plant from which the cuttings are to be made. This will encourage the development of strong, vigorous, fast-grown shoots that have the required high capacity to produce roots.

Propagate evergreen cuttings during late summer to early autumn; rooting will normally take place during winter. However, evergreen plants can be propagated from softer wood earlier in the growing season. Treat these cuttings according to the condition (soft wood, green wood or semi-ripe wood) of their stems.

Prepare the soil in a cold frame by digging thoroughly and mixing in grit, peat and sand.

Evergreen cuttings taken in late summer should be 4–6 in long, although their length must be related to the normal size of the plant and the amount of annual growth. Cuttings from dwarf *Hebe*, for example, may be only 1½ in long. Take a heel with the cutting (see page 124) if it is to be propagated in unsterilized soil in a cold frame or polythene tunnel. Neaten any tail on the heel.

Leave on the cutting any terminal bud that may have set. If, however, growth is continuing, cut out the soft tip with a knife.

Strip any leaves off the bottom third to half of the cutting. Make a shallow vertical wound about 1 in long in the bottom of the stem of plants, such as *Daphne, Elaeagnus* and *Magnolia grandiflora*, that are difficult to root.

Dip the base of the cutting in a rooting hormone powder of ripewood strength (0.8 per cent IBA). Ensure the wound is covered with the powder.

To make more economical use of space, reduce the size of large leaves by cutting off up to half of each leaf-blade with a sharp pair of scissor-type secateurs.

Make a hole with a dibber in the prepared soil in the cold frame and plant the cutting up to its leaves. Allow the leaves of cuttings to touch but not to overlap. If leaves overlap excessively they tend to stick together with a water film, and this provides an ideal place for rotting to develop.

Label the cuttings and water them with a dilute solution of fungicide. Close the lid of the cold frame as tightly as possible. Shade the frame by painting it with a proprietary lime-wash or by covering it with mesh, in order to reduce temperature fluctuation during the day and so prevent scorching. Remove the shading once light intensity and day length decreases during the autumn. If watering is necessary, incorporate a fungicide to help control rots.

Inspect the frame regularly and remove any fallen leaves and dead cuttings. Cover the frame with matting as insulation as long as there is any danger from frost.

Leave the cuttings *in situ* for the whole growing season. Transplant in autumn, taking considerable care when lifting the cuttings as many evergreen plants produce fairly thick, fleshy and brittle roots. Label the new plants clearly.

EVERGREEN CUTTINGS IN A MIST UNIT
Prune the parent plant in winter. In late summer, fill a container with cuttings compost. Then take a cutting about 4–6 in long from the current season's growth. Pinch out any soft growing tip. At the bottom, make a nodal cut.

Strip any leaves off the bottom third of the cutting and make a shallow vertical wound on plants that are difficult to root. Dip the cutting in a rooting hormone.

Cut down the size of large leaves. Make a hole with a dibber in the compost and insert the cutting. Plant any more cuttings, allowing the leaves just to touch. Label and water with a fungicide. Place in a mist unit or closed case.

Ensure the cuttings do not become too wet as over the winter poor light and low temperatures do not dry out composts quickly. Harden them off gradually once the cuttings have rooted. Pot on very carefully in spring and label.

Some plants propagated from evergreen cuttings	Ceanothus	Elaeagnus	Magnolia grandiflora
Abelia	Cherry Laurel	Escallonia	Portuguese laurel
Aucuba	Choisya	Evergreen honeysuckle	Pyracantha
	Daphne	Hebe	

1 Prune the parent plant in winter to encourage strong shoots.

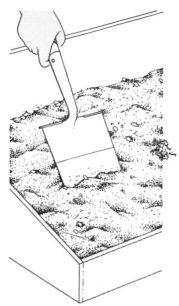

2 Dig the soil in a cold frame thoroughly. Mix in grit, peat and sand.

3 Take a heel cutting in late summer from a stem of the current season's growth.

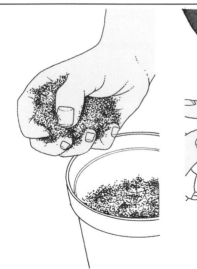

1 Fill a container with cuttings compost. Firm to ³⁄₈ in below the rim.

2 Cut a stem ¹⁄₈ in below a node. Pinch out any tip. Remove the lower leaves.

3 Treat the stem with hormone and plant it. Place in a mist unit.

Evergreens 2

4 Trim the heel. Pinch out any soft growing tip on the cutting.

5 Remove any leaves from the bottom third to half of the cutting.

6 Make a shallow 1 in wound at the base of stems that are difficult to root.

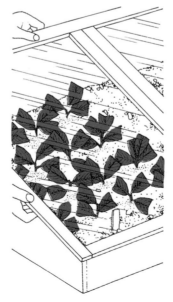

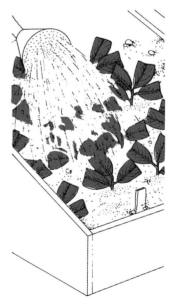

10 Plant any other cuttings. Label them. Water the cuttings with a fungicide.

11 Seal the lid of the cold frame and shade it until light intensity decreases.

12 Prevent the cuttings drying out. Water them with a fungicidal solution.

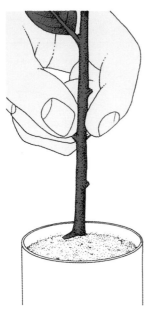

7 Dip the basal cut surface in a rooting hormone. Ensure the wound is covered.

8 Cut down the size of large leaves so they do not overlap.

9 Dibble in the cutting up to its leaves in the prepared cold-frame soil.

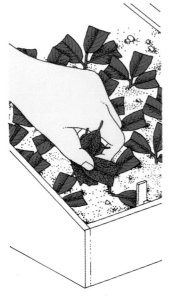

13 Remove any fallen leaves within the frame. Cover with matting as insulation.

14 Harden off the cuttings gradually in spring.

15 Transplant in autumn. Take care when lifting the brittle, fleshy roots.

Hard woods 1

One of the easiest techniques of vegetative propagation is to propagate plants from hardwood cuttings.

A hardwood cutting is made during the dormant season from the fully mature stem of a deciduous tree or shrub. Because the cutting has no leaves, the degree of environmental control required for successful propagation is minimal.

As with virtually all methods of vegetative propagation, it is the preparation of the parent plant, by pruning rigorously a year before the cutting is to be taken, that is possibly the single most important factor in the ultimate success of rooting a hardwood cutting. Hard, rigorous pruning will encourage stems with a high capability of producing roots.

Where and when to take a cutting

A stem grows at different speeds throughout the season. It develops fastest in spring, its rate slowly declining until autumn, when growth ceases altogether. Even by the end of the growing season, the base of a stem that was produced in spring still has the greatest ability to develop roots, and it should be used for most hardwood cuttings.

Plants, such as willows, poplars and currants, that root easily show very little decline in their ability to produce roots anywhere along the stem and so virtually any part of the stem can be made into cuttings.

With plants, such as coloured-leaf plums, that are difficult to root include the swollen base in the cutting, which should be cut flush with the stem (see page 17).

Hardwood cuttings can be taken any time during the dormant season, but they will be most successful at "leaf-fall" and just before the leaf-buds break; their lowest capacity to root is in the early new year.

Cuttings made just before the dormant buds break will need a protected environ-

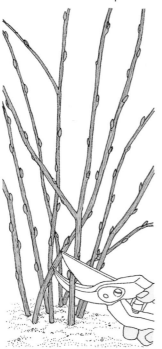

1 Prune the parent plant rigorously during the dormant season.

2 Run a hand down a leafy stem on the parent plant in early autumn.

3 Remove a hardwood stem with all its current year's growth if the leaves fall off.

Some plants propagated from normal hardwood cuttings
Black currant
Cotoneaster – large growing
  hybrids
Currant, flowering
Dogwood, coloured-bark
  (*Cornus*)

Plum, coloured-leaf (*Prunus*)
Poplar (*Populus*)
Rosa – both species and
  modern hybrids
Spiraea – species and varieties
Viburnum – winter-flowering,
  deciduous types
Willow (*Salix*)

ment, such as a cold frame, or very careful timing to avoid leaves being produced before roots, so causing the cutting to lose water too quickly and die. Thus it is more satisfactory to take cuttings at "leaf-fall", when they can be planted in the open ground.

Conventional "leaf-fall" occurs when the stem, having produced a corky abscission layer to isolate each leaf, has its leaves removed by rain, frost or wind. However, as far as the plant is concerned, a leaf is isolated as soon as the corky abscission layer is complete, and this in effect is "leaf-fall".

Run a hand down the stem of a plant. If the leaves fall off, the corky abscission layer is complete, and the time, therefore, is right to take hardwood cuttings.

Size of a hardwood cutting

Although traditionally hardwood cuttings are made between 10–14 in long, a shorter length is much more successful.

Hardwood cuttings, although leafless, will still lose some water by evaporation from their surface. The commonest reason why these cuttings may fail to develop roots is because they are allowed to dry out. To avoid water loss, expose as little of the cutting as possible above the ground. However, if the cutting is planted too deep, the buds will not grow properly. Thus it is vital to expose sufficient of the cutting above ground for about three buds to develop. In practice the third bud can be planted just below ground level as at that depth its growth will not be inhibited. Therefore, for most plants, only 1 in or so of the cutting need be above ground.

A cutting initially develops roots both along the stem and from the cut area at its base. Gradually, the roots along the stem disappear and the root system of the new plant develops from the basal roots alone. Therefore these basal roots should be encouraged by applying a rooting hormone to the cutting and by

4 Make a sloping cut just above the proposed top bud.

5 Make a horizontal cut exactly 6 in below the top cut.

6 Dip only the basal cut in rooting hormone.

Hard woods 2

maintaining good aeration around it. The surface layers of the soil, that is about the top 2 in, have most air in them, so ideally a cutting should be planted within these layers.

However, a cutting planted 2 in deep with about 1 in exposed above ground will be too short for survival: the cutting will dry out too easily; it will be too short to anchor rigidly; and it will be too small to contain sufficient food reserves to support it through the dormant season. Thus the length must be increased to take account of these factors, and 6 in is a satisfactory compromise for a hardwood cutting.

Propagating a hardwood cutting

Prune the parent plant rigorously during the dormant season to encourage fast-growing stems.

At "leaf-fall" cut a hardwood stem with all the current year's growth, using a pair of secateurs. Cut it flush with the parent stem to avoid snags. Make a sloping cut as close as possible above the proposed top bud, remembering that the bottom part of the cutting has the greatest rooting capacity. Make a horizontal cut exactly 6 in below the top cut ignoring the position of any buds.

Treat only the cut base of the cutting, and not the stem, with a rooting hormone powder. The hardwood strength is 0.8 per cent IBA. If suitable stems are made into cuttings at the correct season then the rooting hormone will have little effect except with plants that are difficult to root.

Bundle the cuttings made at "leaf-fall" into quantities of ten or twelve and heel them into a sand-box, almost to their full depth; label, and leave for the rest of the winter. Alternatively, they can be planted directly into well-cultivated soil.

Easily rooted plants will readily survive in the open ground, but less easily rooted plants will benefit from the increased temperatures

7 Heel in some bundles of cuttings in a sand-box and label them.

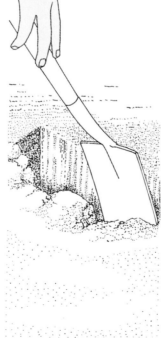

8 Dig the soil very well just before bud-break. Make a furrow 5 in deep.

9 Lift the cuttings. Plant them vertically in the trench 4–6 in apart.

and less fluctuating environment provided by a cold frame.

Just before the dormant buds break in spring, dig the propagation bed thoroughly. Make a furrow 5 in deep with a spade. Lift the cuttings from the sand-box and plant them vertically in the furrow. Allow 4–6 in between cuttings; leave 12–15 in between rows in the open ground and only 4 in between rows in a cold frame. Firm back the soil, leaving about 1 in of the cuttings exposed. Label each row.

Open-ground cuttings may need re-firming if lifted by frost. Leave the cuttings *in situ* for the growing season. In autumn, lift the rooted cuttings, each of which will have produced several stems, and transplant them to their final situations in the garden.

Protect cuttings planted in a cold frame only until they have developed some roots and then harden off. It becomes increasingly difficult to harden them off as the growing season progresses.

SINGLE STEMS

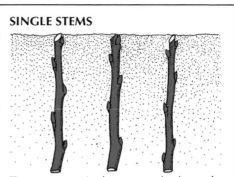

To grow a single-stemmed plant, for example a tree, rather than one with several stems, encourage only the top bud to develop on the cutting. Plant the cutting vertically in the soil so that the tip is just covered. This will inhibit the development of the lower buds and so produce a single stem.

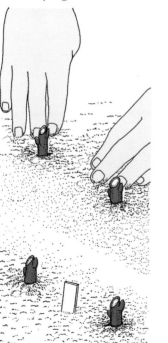

10 Firm back the soil, leaving about 1 in of the cuttings exposed. Label.

11 Re-firm the cuttings if they are lifted by frost.

12 Lift the rooted cuttings in the following autumn. Transplant and label them.

Hard woods 3

Disbudding

A plant is sometimes needed that has a single stem, or "leg", at the bottom and many stems, or branches, farther up the plant. To grow a plant with this characteristic, a hardwood cutting needs to be longer than the normal 6 in and have more than 1 in exposed above the ground. However, a longer cutting may encourage further buds to develop lower down the stem, so losing the desired single stem.

To prevent any branching down the stem, cut out all but the top three buds at "leaf-fall". Cut out the buds shallowly but completely with a sharp knife; ensure no latent buds or part-buds remain. It is simpler to disbud in this way than to cut out stems at a later date.

Gooseberry and red-currant bushes are normally propagated from 10–14 in hardwood cuttings that have been disbudded. The bush will then have a single stem at the bottom, which will allow free air circulation around the bush, thus reducing the possibility of mildew attacks.

To produce rose rootstocks take an 8 in hardwood cutting and remove all but the top two buds from plants such as *Rosa laxa* and *Rosa multiflora* 'Simplex'. This method will avoid rootstock suckers.

Seal the areas that have been disbudded to prevent rotting and disease. It is possible to leave them to callus over naturally, but unless this is done under fairly humid conditions there is always the danger that the cutting may desiccate.

It is better, therefore, to paint the disbudded areas with candle wax.

Treat the cuttings with rooting hormone. Plant the rootstock and label it. For roses, leave the two top buds exposed and sufficient stem above the soil level for any buds to be grafted on later in the season.

Gooseberry and red-currant cuttings should be planted with their third top bud within 2 in of the soil surface. At the end of the growing season lift the cuttings and replant with much more of the stem exposed.

Soft- or hollow-pith stems

For many woody plants, such as *Forsythia*, it is not possible to propagate hardwood cuttings by the method described on pages 138–41 because they have a soft or hollow pith. If this is exposed, it very often provides a site for rots and diseases, which will then kill the cuttings.

There are two ways of overcoming this problem: either make the hardwood cutting exactly 6 in long and then seal the base of the cutting with candle wax, or make the basal cut below a leaf-bud joint. Cuttings treated by either of these methods should root just as prolifically as cuttings with a solid pith.

Melt some candle (paraffin) wax until it is liquid. Touch the base of a 6 in cutting in the wax so that a drop of wax attaches, cools and sets quickly, so sealing the cut. Do not overheat the wax and so damage the plant.

This is a very satisfactory method of sealing these cuttings, but take care not to damage the wax seal when the cuttings are bundled, heeled in and planted.

The alternative method is to make the basal cut at a node, that is immediately below a leaf joint, where the pith is generally solid. Make the basal cut at the node nearest to the normal 6 in cutting length. Plants such as *Kerria* very often have long spaces between nodes, and cuttings may then become as much as 8–9 in long.

It is often recommended that plants with a soft pith should be cut with a knife, as this tends to cause less damage than secateurs, which may crush the cutting. Provided, however, that a sharp pair of scissor-type secateurs is used, it is rare for extensive damage to occur and secateurs are, therefore, quite satisfactory.

Some plants that may need to be disbudded
Gooseberry
Red currant
Rosa laxa
R. multiflora 'Simplex'

Some plants propagated from hardwood cuttings having a soft or hollow pith
Deutzia
Forsythia
Kerria
Leycesteria
Philadelphus
Weigela

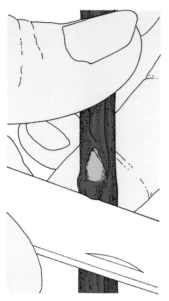

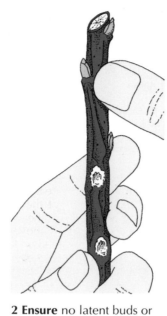

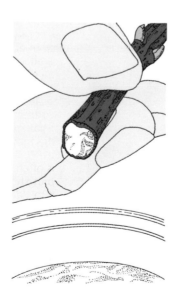

1 Cut out all but the top three buds on a cutting at "leaf-fall".

2 Ensure no latent buds or part-buds remain on the cutting.

3 Paint the disbudded areas with wax or a pruning paint.

Sealing an internodal cut

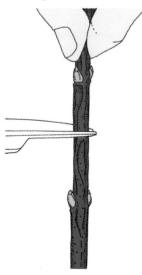

1 Take a cutting exactly 6 in long. Melt some candle wax until it is liquid.

2 Dip the base of the cutting in the wax to seal it.

Making a nodal cut

Make a cut just below the leaf joint that is 6 in or more from the stem top.

143

Hard woods 4

Vine-eye cuttings

This technique is normally used only for vines, although it is quite possible to adopt it for any plant with a reasonably solid pith that can be propagated by hardwood cuttings. Vine-eye cuttings are the hardwood equivalent of leaf-bud cuttings.

Prune back the plant during the dormant season to encourage vigorous growth.

Fill a small pot with cuttings compost and firm to within ³/₈ in of the rim, using a presser board.

At "leaf-fall" cut a stem with all the current season's growth from the parent plant. Make a sloping cut with a pair of secateurs just above a leaf-bud joint, so that no snag is left that could die back and possibly kill the bud. Make the basal cut horizontally across the stem about 1½ in below the top cut. Wound the cutting if the plant is difficult to root. Place a sharp knife half-way down the stem on the opposite side to the bud. Make a shallow cut down to the base.

Dip the basal cut surface and the wound into hormone powder of a suitable strength for hardwood cuttings (0.8 per cent IBA). Make a hole in the prepared compost with a dibber. Plant the cutting vertically so that the bud lies about level with the compost surface. Insert only one cutting per pot.

Label the pot and stand it on a greenhouse bench or in a closed case – the higher the temperature, the faster will be the rate of regeneration.

Water the cutting to prevent it drying out. Do not overwater during the winter when the cutting is dormant, as the compost will readily waterlog, causing the cutting to rot and die. Harden off the cutting once it has rooted, and transplant in spring. Label it.

4 Make a cut horizontally across the stem 1½ in below the top cut.

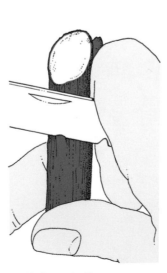

5 Make a shallow cut on the side opposite the bud from half-way down the stem to the base.

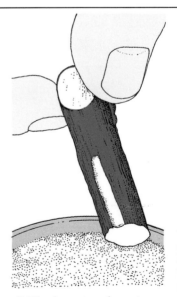

6 Dip the cut surfaces in rooting hormone. Make a hole with a dibber in the prepared compost.

1 Prune back the parent plant during the dormant season to encourage strong, vigorous stems.

2 Fill a small pot with cuttings compost. Firm to within ³/₈ in of the rim, using a presser board.

3 Cut a stem with all the current season's growth at "leaf-fall". Make a sloping cut just above a leaf-bud.

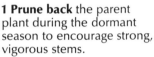

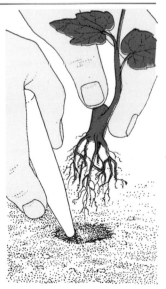

7 Plant the cutting vertically so that the bud lies about level with the compost surface. Label it.

8 Leave the potted cutting on a greenhouse bench. Water to prevent the cutting from drying out.

9 Harden off the cutting as soon as it has rooted. Transplant in spring.

Conifers 1

CONIFER SHOOTS

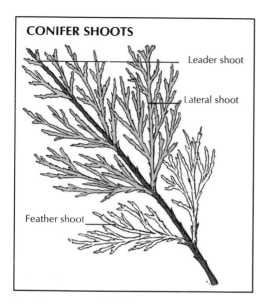

Leader shoot

Lateral shoot

Feather shoot

Conifers are predominantly evergreen trees and shrubs, some of which can be propagated from softwood, greenwood, semi-ripe and ripewood cuttings. In general, spruces, firs and pines do not grow well from cuttings, and they should be propagated from seed or, if special varieties are required, by grafting.

Whether conifer cuttings are to be propagated in a warm environment, such as a closed case, or in a colder one, such as a cold frame, does not affect when they are taken.

Take cypress cuttings in autumn or winter, and yew and juniper cuttings in the new year after the parent plants have been subjected to a period of frost.

Select cuttings from young, actively growing plants that are clipped regularly, and so produce strong, vigorous shoots.

Preferably take cuttings from the top rather than the bottom of the plant, as cuttings from low down often develop into plants of

Conifers propagated in a warm environment

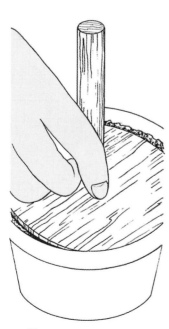

1 Fill a container with some cuttings compost. Firm to within ⅜ in of the rim.

2 Cut off a vigorous leader or lateral shoot about ¼ in into the brown-barked wood.

3 Trim the leaves off the bottom 1–1½ in of the cutting. Dip the basal cut in rooting hormone.

atypical habit in that they may continue to grow horizontally.

Conifers vary considerably in their pattern of growth. There are leader shoots, which are the growing points of the branches; lateral or sub-terminal shoots, which also increase in size; and feather shoots, which do not grow larger.

For propagation by cuttings it is important to choose shoots with a distinct growing point, and this is why feather shoots are unsatisfactory. If a growing point is not present, bun-shaped forms will be produced, particularly with yellow foliage forms.

Conifer propagation in a warm environment
Conifer cuttings to be propagated in the warm environment of a closed case or mist unit should have a softer base than those to be propagated in a cold frame; they should also become established more quickly.

Fill a container with cuttings compost and firm to within ³/₈ in of the rim.

The tip of a conifer shoot is green; it gradually turns yellow further down the stem, and then brown. Remove a vigorously growing leader or lateral shoot from the parent stem. Make a clean cut about ¹/₄ in into the brown-barked wood so that the cutting is predominantly green but has a small protective "plug" of hard wood at its base.

Trim the leaves off the bottom 1–1¹/₂ in of the cutting. Do not remove the growing point. Dip the cutting in a rooting hormone powder of ripewood (0.8 per cent IBA) strength. Make a 1 in hole with a dibber in the compost and insert the cutting and firm. Leave about 1¹/₂–2 in between cuttings. Label the container; then water with a fine rose. Place in a closed case or mist unit.

The cuttings will root within three to four months. Harden off and pot on in spring.

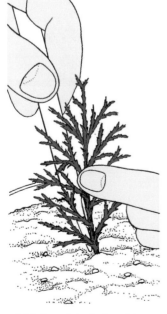  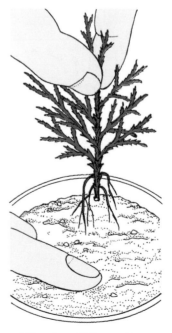

4 Make a 1 in hole with a dibber. Insert the cutting and firm. Plant any other cuttings 1¹/₂–2 in apart.

5 Label them. Then water with a fine rose. Place the container in a closed case or mist unit.

6 Harden off the cuttings once they have rooted. Pot on in spring. Label them.

147

Conifers 2

Conifers propagated in a cold frame
It is easier, but slower, to propagate conifer cuttings in a cold frame.

Prepare the soil in the propagation bed before taking the cuttings. Raise the level of the soil in the cold frame to within 6–8 in of the lid. This will maintain an equable humidity and temperature for the cuttings.

Pull a stem backwards down the parent stem to obtain a cutting with a heel, which can then be propagated in a cold frame.

Trim the heel of the cutting with a sharp knife. Remove the leaves on the bottom third of the cutting either with a knife or by hand provided only a small scar is caused which may actually encourage the cutting to root. Do not remove the growing point.

Dip the base of the cutting in a rooting hormone powder of ripewood strength (0.8 per cent IBA). Make a 1 in hole with a dibber in the cold-frame soil and plant a cutting. Firm the soil back round the stem. Space any other cuttings 2 in apart and label them. Water with dilute fungicidal solution and close down the lid of the cold frame.

Ensure the environment is always kept absolutely clean and hygienic as the cuttings have to survive for nearly a year in the cold frame. Place matting on the lid to insulate the cuttings during any cold weather. Remove the matting as soon as the danger of frosts has passed.

Leave the cuttings undisturbed until summer, when the frame should be aired and shaded to prevent scorching. This can be done either by painting its lid with a proprietary brand of lime-wash or by covering it with mesh. Water the cuttings as necessary to prevent them drying out.

Lift the cuttings in autumn. Pot on or transplant to their final position in the garden.

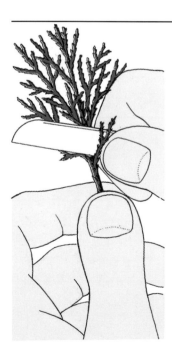

4 Remove the leaves on the bottom third of the cutting with a knife or by hand.

5 Dip the basal cut surface in a rooting hormone powder.

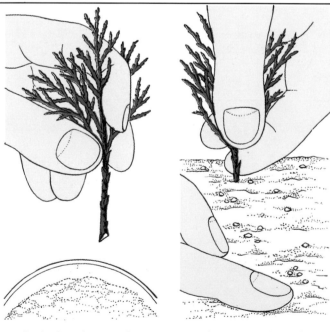

6 Make a 1 in hole in the prepared cold-frame soil. Insert the cutting and firm the soil around it.

Conifers propagated in a cold frame

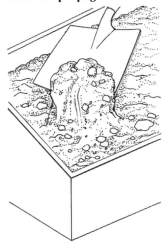

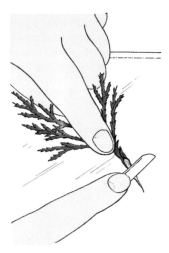

1 Raise the level of the soil in the cold frame to within 6–8 in of the lid.

2 Take a heel cutting from a strong, vigorous leader or lateral shoot.

3 Neaten the long, straggly tail on the heel of the cutting.

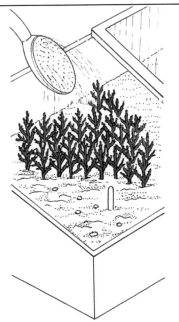

7 Plant any other cuttings 2 in apart. Label; then water with a fungicide. Close the cold frame.

8 Insulate the cold frame with matting until the danger from frosts has passed.

9 Shade the frame from intense light in summer. Transplant the rooted cuttings in autumn.

Sub-shrubs

Some woody, low-growing plants, such as *Salvia*, that are often treated as herbaceous plants are called sub-shrubs. Most of them root readily and can easily be propagated from stem cuttings taken in late summer.

Prune the parent plant during the dormant season to encourage strong, vigorous shoots.

Prepare the soil in a cold frame by digging deeply. Add grit if necessary to improve drainage.

Cut off some shoots with all their current season's growth in late summer (usually early September). Select non-flowering shoots if possible, although with many plants, such as lavender, these may be difficult to find. If flowering shoots have to be used then cut back the flower and its stem to the leafy part.

Cut out any soft growing tip with a sharp knife or pinch it out between the thumb and forefinger. Make the basal cut with secateurs about 4 in below the top of the cutting. A quick and easy way to measure 4 in is to hold the stem in the palm of an adult's hand. The stem will be approximately 4 in long where it reaches the butt of the hand.

1 Prune the parent plant during the dormant season to encourage strong shoots.

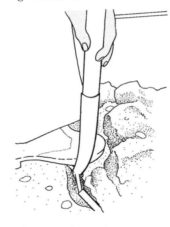

2 Prepare the soil in a cold frame by digging deeply. Add grit if necessary.

3 Collect a non-flowering shoot in late-summer. Cut out any soft growing tip.

7 Make a hole and plant about half the cutting in the prepared soil.

8 Plant the other cuttings about 4 in apart. Label and water with a fungicide.

9 Seal the cold frame and shade it until the light intensity decreases.

Strip the leaves from the bottom half of the cutting, either by pulling very carefully or by cutting with a sharp knife. Dip the base of the cutting into a rooting hormone powder of semi-ripewood strength (0.4–0.5 per cent IBA).

With a dibber make a hole in the prepared soil about half the length of the cutting, that is about 2 in deep. Firm the soil around the cutting. Leave about 4 in between each cutting and about 4 in between each row.

Label the cuttings clearly. Using a watering can with a coarse rose, apply a fungicide over the cuttings to protect them against disease.

Seal the cold frame and shade it to avoid scorching. Check the cuttings regularly; water when necessary to prevent them drying out. Remove the shading on the cold frame and reduce watering as soon as light intensity decreases. Insulate the cuttings against frost by laying some matting over the cold frame. Remove this during the day, if possible.

Harden off the rooted cuttings gradually in spring. Then lift and transplant to their final position in the garden. Label them clearly.

4 Make a basal cut with secateurs about 4 in from the top of the cutting.

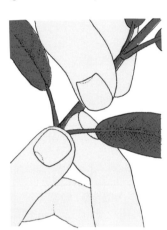

5 Trim the leaves from the bottom half of the cutting.

6 Treat the bottom of the cutting with a rooting hormone powder.

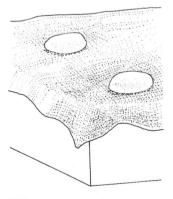

10 Lay some matting over the frame to insulate the cuttings against frost.

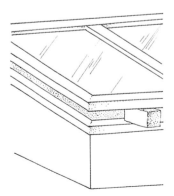

11 Harden off the rooted cuttings gradually in spring.

12 Lift and transplant them once they are accustomed to the climate. Label them.

Rhododendrons

The genus *Rhododendron* is an extremely large and varied group of plants that have differing abilities to root from cuttings. The large-flowering hardy hybrids and the small-leafed dwarf rhododendrons are the most worthwhile to propagate from stem cuttings.

Large-flowered hardy hybrids

Prepare the compost before taking any cuttings to avoid unnecessary delay, and water loss, before planting. Mix together equal parts lime-free grit and sifted peat. Fill a container with the compost and firm to within ³/₈ in of the rim.

Take cuttings from the second flush of growth at the end of the summer, usually in September. The second flush of growth can be recognized by the umbrella of shoots that does not occur on the first flush of growth.

Remove all the leaves except the terminal whorl (the group of leaves nearest the tip). Snap out the terminal bud regardless of whether it is a flower or vegetative bud.

Place the cutting in the palm of a hand so that the top is just about level with the forefinger and the stem is across the palm of the hand. Cut the stem at the butt of the hand with a sharp pair or secateurs. This will give a cutting of about 4 in long. Cut part of the remaining leaves to reduce their leaf area and so make planting easier.

Wound the bottom ³/₄ in of the cutting by making a shallow score with a sharp knife. Dip the cut surfaces in a strong rooting hormone (0.8 per cent IBA). This is particularly important for varieties with 'Britannia' in their pedigree, as they are generally difficult to root.

Make a hole with a dibber in the prepared compost and insert the cutting. Plant any further cuttings as close together as possible.

Label the container and place it in a well-lit environment with bottom heat of 21°C/70°F. Control water loss by covering the container with a very thin, clear polythene sheet or by leaving the container in a mist unit.

Prevent rots by applying a dilute fungicidal solution at regular weekly intervals.

Rooting should not be expected until well into the new year.

Harden off the rooted cuttings and transplant in late winter/early spring and label. Place them in a cold frame to grow on.

1 Fill a container with equal parts lime-free grit and sifted peat. Cut a stem of the second flush of growth in late summer.

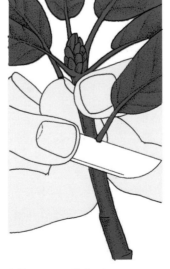

2 Remove all the leaves except the terminal whorl. Snap out the terminal bud, irrespective of it being a flower or vegetative bud.

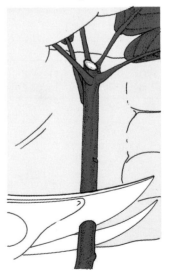

3 Reduce the cutting to 4 in, using a sharp pair of secateurs. Cut off part of the remaining leaves to make planting easier.

DWARF RHODODENDRONS

These are the easiest group from which to take cuttings, although some varieties, depending on their parentage, are more difficult than others to root successfully.

Mix thoroughly equal parts lime-free grit and sifted peat. Fill a container of suitable size with compost – allow about 1½ sq. in per cutting – and firm gently.

Select a stem of the current year's growth that has set its terminal bud, which is usually in late August to September. Cut back the stem with a sharp pair of secateurs to about 1½ in from the tip. Pick or cut off any leaves on the bottom half of the cutting.

Wound the cutting by making a very shallow slice on the bottom ½ in of the cutting. This wounding is not always necessary, but many varieties do benefit, and it is advisable to adopt a standard procedure. Treat the cut surfaces with rooting hormone powder, using the hardwood strength of 0.8 per cent IBA.

Make a ½ in hole with a dibber in the prepared compost. Plant the cutting and water with a fungicide. Label; then place in a well-lit humid environment with bottom heat; a mist unit is ideal to produce these conditions. Once the cutting has rooted, harden off gradually in spring. Pot up and label.

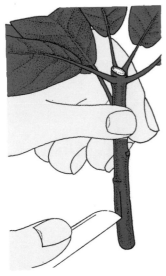

4 Make a shallow score in the bottom ¾ in of the cutting. Dip the cut surfaces in a rooting hormone.

5 Dibble the cutting into the compost. Label the container and place it in a mist unit with bottom heat of 21°C/70°F.

6 Spray each week with a fungicide. Harden off the rooted cutting and repot in early spring. Label and place in a cold frame.

153

Heathers

An extremely prolific and relatively easy way to propagate heathers is from cuttings.

The season to do this is summer, although the exact timing depends upon the availability of non-flowering shoots from which to make the cuttings.

Do not take cuttings from shoots with flower buds that have set because rooting may well be slow and poor, especially for winter-flowering heathers. However summer-flowering heathers can be propagated from shoots that have already flowered, should non-flowering shoots be scarce.

Prune a plant during the dormant season to encourage strong, vigorous shoots.

Make up some cuttings compost of equal volumes lime-free grit and sifted sphagnum moss peat. It is imperative that the peat is sifted and all lumps removed, otherwise later on the rooted cuttings will be difficult to separate without damage.

Choose a container appropriate to the number of cuttings to be taken, allowing ¾ in between cuttings. A separate container should be used for each variety as varieties will root at different rates.

Fill the container with the compost and firm with a presser board to within ⅜ in of the container rim.

Cut off a 1–1½ in vegetative shoot with a pair of sharp scissors. Remove the leaves on the lower part of the cutting. The cutting is then ready to be planted as there is no need to apply a rooting hormone.

Make a hole with a thin dibber to half the depth of the cutting, which is then inserted. Plant any remaining cuttings ¾ in apart. Label; then water with fungicide, using a fine rose. Do not firm by hand.

Place the container of cuttings in a protected environment. Quickest rooting will occur if there is bottom heat and high humidity, for example in a mist unit. A well-sealed cold frame, shaded in the summer and insulated with matting in the winter, will suit just as well, although rooting of late-season cuttings may not occur until the spring.

Harden off the rooted cuttings and pot on in spring. Take special care when knocking out the cuttings and teasing them apart, so that minimal damage is done to the roots. Label the new plants clearly.

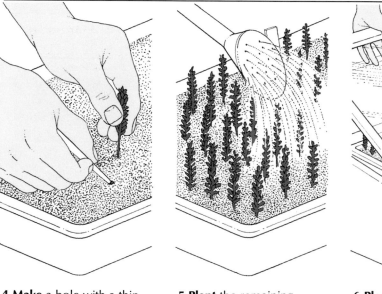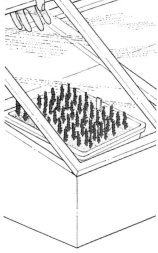

4 Make a hole with a thin dibber to half the depth of the cutting, which is then planted.

5 Plant the remaining cuttings ¾ in apart. Label; then water with a fungicide using a fine rose.

6 Place the container of cuttings in a protected environment.

1 Prune a plant during the dormant season to induce vigorous growth.

2 Mix equal parts lime-free grit and sifted sphagnum moss peat and place in a container.

3 Cut off a 1–1½ in vegetative shoot. Remove the leaves on the lower part of the cutting.

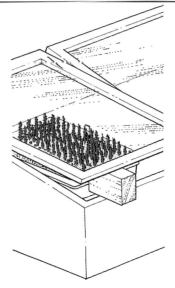

7 Harden off the cuttings gradually once they have rooted.

8 Pot on in spring. Take special care of the roots, which are easily damaged.

WHEN TO TAKE CUTTINGS

late May to early July
Erica × darleyensis
E. erigena (mediterranea)
E. carnea (herbacea)

late June to early August
Erica vagans

July to August
Erica arborea
E. australis
E. lusitanica
E. × veitchii

August to September
Daboecia cantabrica
Erica ciliaris
E. cinerea
E. tetralix
E. × watsonii

late August to early October
Calluna vulgaris

Leaves

Some plants – mostly house plants that belong to *Begoniaceae, Crassulaceae* and *Gesneriaceae* – have the capacity to develop plantlets on their leaves. This is a simple and efficient means of propagation, and it can occur in either of two ways: by naturally growing foliar embryos, or by artificially induced plantlets from leaf cuttings.

Foliar embryos are the result of a highly specialized process that occurs in certain plants, such as the mother-of-thousands and pig-a-back plant. In this process the plant isolates simple plant cells in small areas of its leaves during the course of its growth; these cells are subsequently capable of developing into new plants.

The range of plants that can be propagated from leaf cuttings is relatively small, and the success of this technique is subject to various environmental factors that are to some extent open to manipulation by the gardener.

Leaf cuttings should be made only from leaves that have recently expanded fully. If the leaf is still less than full size and immature, all its energy will first go towards developing and maturing. This will delay the generation of new plant life and, since a leaf cut off from its parent is unsupported, the longer the propagation process takes the more likely it is that problems, such as rotting, will arise.

When a leaf has recently expanded to its maximum leaf area it is efficient in food production and still has a full life expectancy in case regeneration should be slow. At this time the leaf is still young enough to have a high capacity to propagate – a capacity that will lessen as the leaf ages.

The selected leaf should be complete, normal and undamaged so that it will not be

Asplenium bulbiferum (Chicken fern)

Tiarella cordifolia (Foam flower)

Saintpaulia ionantha (African violet)

subject to rots and will produce typical off-spring. It should also be free from pests and diseases.

Since most plants suitable for propagation by leaf cuttings are grown indoors or under glass, it is possible to take cuttings all year round, as long as there is a fresh, fully expanded leaf available. All that will alter is the speed at which the plantlets develop; in winter, temperatures and light intensity will be lower and, as a consequence, food production and the rate of propagation will be slower.

The leaf, when separated from its parent plant, will be highly susceptible to desiccation and it is necessary to minimize this by controlling the environment. Therefore always propagate leaf cuttings in a closed case, propagator, or under a glass sheet or polythene tent.

The most common cause of failure in leaf propagation is the leaf rotting before it has a chance to produce a self-supporting plant. Thus it is important that all materials, containers, composts and leaves should be clean and undamaged.

Although many plants have leaves that are capable of rooting, they do not all have the capacity to propagate from leaf cuttings. Those that do not can only be successfully propagated from leaf-bud cuttings (see pages 122–3). Other specialized kinds of leaf cuttings are bulb scaling (see pages 94–5), and scooping and scoring bulbs (see pages 96–7).

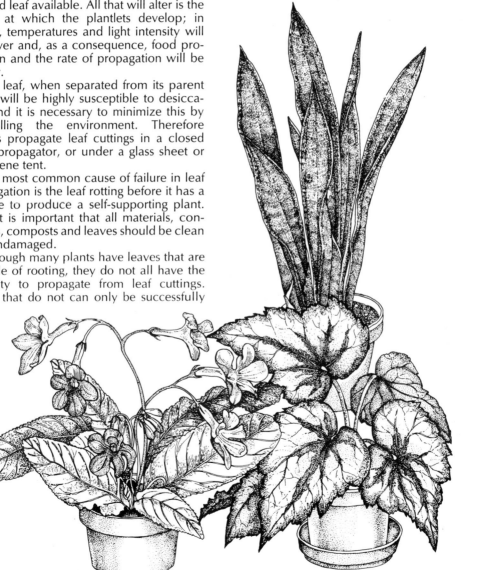

Sansevieria trifasciata
(Mother-in-law's tongue)

Streptocarpus × *hybridus* (Cape primrose)

Begonia rex

Leaf-petiole cuttings

The simplest and most reliable way to produce new plants from leaf cuttings is to use a complete leaf with its stalk. The disadvantage of this method is that it develops only a few new plants from each leaf.

Rotting and disease are the main causes of failure so always use clean tools, containers and composts.

Leaf-petiole cuttings can be taken at any time of year provided a new, fully expanded leaf is available.

Make up a cuttings compost of equal parts sifted peat and grit. Fill a container that is large enough to take the leaf-petiole cuttings. Firm the compost to within ⅜ in of the rim of the container.

Slice a suitable leaf cleanly through the leaf-stalk, using a sharp knife or a safety razor blade to ensure the least possible damage. Leave about 2 in of the leaf-stalk attached to the leaf-blade.

Make a small hole with a dibber in the compost to a depth just sufficient to hold the cutting. Plant at a shallow angle so that the leaf-blade is almost flat on the compost. The shallower the base of the stalk is planted into the compost the more the air can circulate around it, which will encourage a quick response. Then firm the compost around the stalk. When the cuttings are all planted, label them and water in using a fine rose to avoid damaging them.

Place the cuttings in an environment that maintains a steady high humidity, so that the cuttings do not dry out. The temperature, especially for house plants, needs to be relatively high, and this is best provided by using a propagator that is heated at the bottom – ideally at about 20°C/68°F.

Expose the cuttings to sufficient light for them to manufacture food and develop the new plantlets. Too much sunlight may scorch the cuttings in general, light shade is the best compromise.

The new plantlets will develop on the cut surface of the leaf-stalk within five to six weeks, and several may appear at this point. Leave them until they can be handled and separated into individual plants, potted on and hardened off. If it is likely to be some time before they are large enough to be potted on, then liquid feed the plantlets.

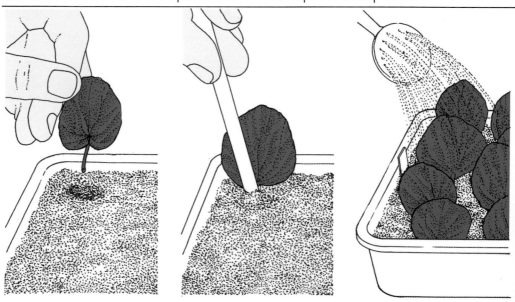

4 Make a shallow hole and insert the cutting at the angle shown.

5 Firm the compost gently round the cutting. Plant the remaining cuttings.

6 Label and water in, using a fine rose.

Some plants responding to propagation by leaf-petiole cuttings
Begonias – other than *Begonia rex*
Peperomia caperata
P. metallica
Saintpaulia

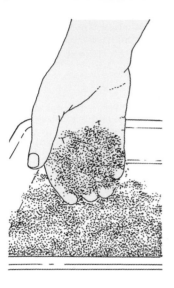

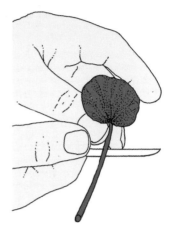

1 Fill a container with equal parts sifted peat and grit.

2 Slice an undamaged leaf that has recently expanded fully away from the plant.

3 Cut through the leaf-stalk about 2 in from leaf-blade, using a sharp knife.

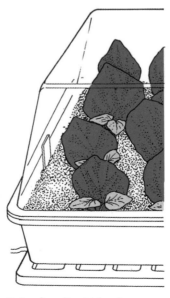

7 Place the cuttings in light shade in a propagator that is heated from below.

8 Apply a liquid feed once the new plantlets have started developing.

9 Pot on the plantlets once they can be handled. Label and harden them off.

159

Midrib lateral vein cuttings

A leaf-midrib is the extension of a leaf-stalk, and it is possible to propagate from leaf-midrib cuttings in the same way as it is from leaf-stalk (petiole) cuttings.

Leaf-midrib cuttings can be used for any leaves that have a single central vein, and it is a particularly successful technique with *Streptocarpus*.

Before taking cuttings, ensure all tools and equipment are scrupulously clean. Disease can be a major problem if care is not taken when propagating.

Leaf-midrib cuttings can be taken at any time of the year provided the plant has a suitable leaf, that is one that is undamaged and has recently expanded fully.

Fill a plastic container with cuttings compost. Water thoroughly and allow to drain.

Cut a suitable undamaged leaf cleanly from the parent plant and place it upside down on a clean sheet of glass. Cut off a strip that is not more than 2 in wide, using a safety razor blade at right angles to the midrib. Cut further strips in the same way, each strip having a central rib and two wings.

Make a shallow trench in the compost and insert the bottom of a cutting just deep enough to hold it erect. Firm gently. Plant the remaining cuttings about 1 in apart.

Spray the completed container with a fungicide to protect the cuttings against fungal rots.

Label the container and place it in a warm (21°C/70°F), humid environment to encourage the cuttings to root. However, ensure there is sufficient light for the leaves to manufacture

---

LEAF CUTTINGS ON LATERAL VEINS

Lie a leaf upside down on a clean sheet of glass. Remove the midrib with a razor blade so that the two halves of the leaf-blade are isolated and all the lateral veins have an exposed cut surface.

Make a shallow trench in a container filled with moist cuttings compost. Plant the leaf cuttings vertically in the trench with the cut surfaces of the lateral veins just in the compost. Firm gently; label and place in a propagator or closed case.

Plantlets will develop on the cut surfaces of the lateral veins within five to eight weeks. Separate, pot on and label when they are large enough to handle.

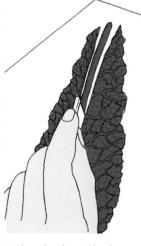

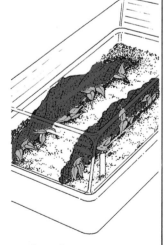

1 Lie a leaf upside down on a clean sheet of glass. Remove the midrib with a razor blade.

2 Make a shallow trench in some moist cuttings compost. Plant the cuttings upright. Firm and label.

3 Place the container in a propagator. Pot on the plantlets once they can be handled. Label clearly.

food. Avoid direct sunlight, which may scorch the cuttings. A propagator or a closed case with bottom heat is ideal, although a polythene tent supported with a cane or loop of wire is quite adequate protection.

Stand the container in a water bath to rewet compost that is beginning to dry out.

Young plantlets should appear in five to eight weeks, but they will not be big enough for transplanting for several more weeks.

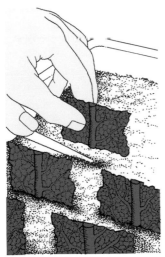

1 Fill a container with cuttings compost. Water thoroughly and allow to drain.

2 Cut an undamaged leaf that has recently expanded fully from a plant.

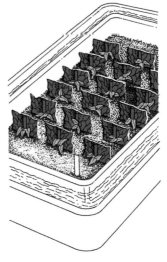

3 Place the leaf upside down on a clean sheet of glass. Cut the leaf into strips not wider than 2 in.

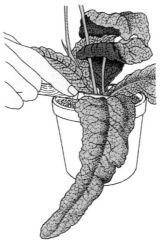

4 Make a shallow trench. Insert a cutting. Firm gently. Plant the remaining cuttings 1 in apart.

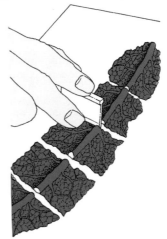

5 Label the container. Spray with a fungicide. Place in a heated propagator.

6 Rewet drying compost by standing the container in water. Pot on the plantlets once they can be handled.

161

Leaf slashing

Plants that do not have a leaf with a central midrib and lateral veins but possess a more netted veining are not easy to propagate from leaf cuttings. Certain plants, however, have leaf veins that are capable of generating a new plant, and these can be propagated by cutting through the veins of the leaf so that plantlet development is induced but the leaf itself remains entire. Because of the size of their leaves this technique is particularly suitable for varieties of *Begonia rex*.

Clean all tools and equipment. Fill a seed tray with cuttings compost; then press it down to about 3/8 in below the rim. Water the seed tray thoroughly and allow to drain.

Remove a fully expanded leaf that is undamaged from the parent plant. Place the leaf upside down on a clean sheet of glass. Cut off the leaf-stalk flush with the leaf-blade, using a safety razor blade. Make a 3/4 in cut across a major vein, using the razor blade. Repeat this until there is one cut every square inch all over the leaf.

Place the cut (or slashed) leaf, with its top side upwards, flat on to the surface of the cuttings compost in a seed tray. Pin the leaf down with a thin wire staple if it does not lie flat on the compost. Label the leaf cuttings clearly. Spray the seed tray with a fungicide to reduce the likelihood of rot, and cover with a sheet of glass.

Place in the light, which is essential for food production and so for plantlet development. Avoid direct sunlight, which may scorch the leaf. A closed case with bottom heat provides the best environment.

The rate at which plantlets develop will be dependent on the surrounding temperature: at 21°C/170°F, plantlets should appear in three to four weeks. Separate the new plants once they can be handled easily and pot on. Label and harden off gradually.

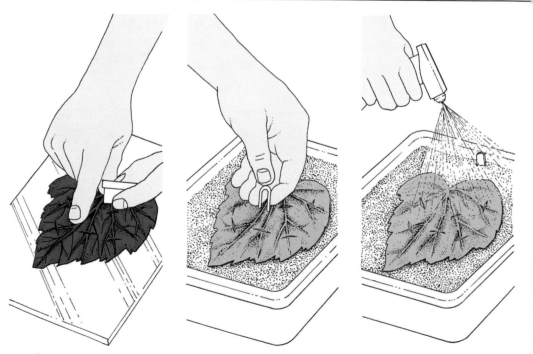

4 Make a 3/4 in cut across a major vein. Repeat this every square inch.

5 Place the leaf top side up on the compost surface. Pin down with a wire staple.

6 Label the seed tray. Then spray with a fungicide.

1 Fill a seed tray with some cuttings compost. Water well and allow to drain.

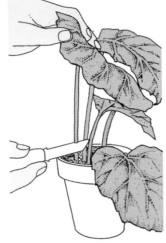

2 Remove a fully expanded leaf that is undamaged from the parent plant.

3 Place the leaf upside down on some glass. Cut off the leaf stalk.

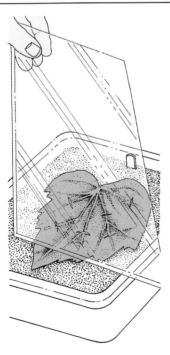

7 Cover with a sheet of clean glass. Place in a warm shaded environment.

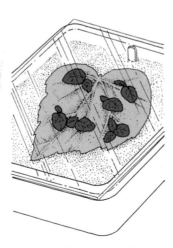

8 Leave the plantlets in their protected place when they first appear.

9 Separate the plantlets once they can be handled. Pot on and harden off.

163

Leaf squares

Any leaf that will regenerate vegetatively can be propagated from leaf squares, but this method is normally used only for plants with big leaves and especially for *Begonia rex* varieties. Its main advantage is that it produces numerous plantlets from a single leaf.

This method of propagation produces small pieces of cut leaf that are prone to rotting. Therefore clean, sterile tools and equipment should be used at all times and scrupulous, hygienic measures should be observed.

Fill a clean plastic seed tray with cuttings compost and press the compost to give a flat surface about ³/₈ in below the rim. Water the compost thoroughly and allow to drain.

Remove a fully expanded, undamaged leaf from the parent plant and place it face down on a clean sheet of glass. Take a safety razor blade and ruler and cut this leaf into a series of ³/₄ in squares, placing the ruler gently on the leaf to avoid crushing. Any damaged squares must be discarded.

Lay the cuttings flat on the compost with the top side facing upwards. Place in rows about ½ in apart. Label the seed tray clearly. Spray the cuttings with a fungicide.

Cover the seed tray with a clean sheet of glass to maintain humidity. Place in a warm (18–21°C/65–70°F) environment out of direct sunlight, but make sure there is sufficient light available to allow the leaves to manufacture food. A closed case is ideal, although a window sill in a warm room facing east or west is also suitable.

If the compost was thoroughly watered initially and the seed tray is covered, little or no more watering will be needed. Should the compost dry out, rewet by standing the seed tray in a water bath.

Plantlets should appear on the cut surfaces of the larger leaf veins nearest to the leaf-stalk. At a temperature of 21°C/70°F, this should occur after about five or six weeks, but the new plants will not be big enough to pot on for several more weeks. When the first leaves have opened, gradually harden off the plantlets by airing the seed tray. Pot up when the plants are a sufficient size to handle without damage.

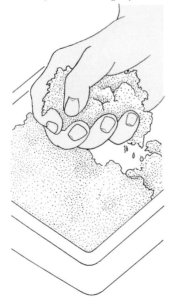

1 Fill a container with cuttings compost. Press to within ³/₈ in of the rim. Water well; then drain.

2 Remove a fully expanded undamaged leaf from a plant. Place it face down on a clean sheet of glass.

3 Cut the leaf into ³/₄ in squares, using a razor blade and ruler.

WRINKLED OR HAIRY LEAVES

Wrinkled, puckered or hairy leaves such as those on *Begonia masoniana* (Iron Cross) will not lie flat in contact with the compost. They should, instead, be planted vertically in the compost, and be just deep enough to hold the leaf squares firmly erect. Ensure that the basal cut, which will eventually produce the plantlets, is in the compost.

Label. Spray with a fungicide; then place in a closed case or a polythene bag.

1 Cut a wrinkled leaf into ³/₄ in squares.

2 Plant vertically in the compost. Firm in well.

3 Spray with a fungicide. Cover with polythene.

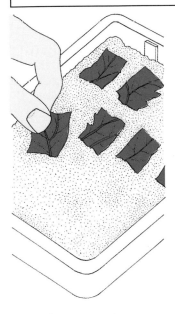

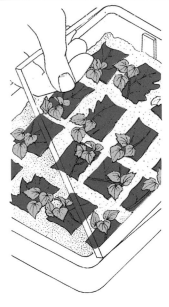

4 Lay the cuttings face up on the compost about ¹/₂ in apart. Label and spray with a fungicide.

5 Cover with some glass. Place in a warm, shaded area. Harden off once the plantlets produce leaves.

6 Pot up when the plantlets are large enough to handle without damage and label.

Monocot leaves

Monocotyledonous plants, such as snow-drops and mother-in-law's tongues, have a series of parallel veins running the length of each leaf. Some of these plants can be propagated at any time of year from leaf cuttings, providing they have an inherent capacity to produce a plantlet on the cut surface of a vein and suitable leaves are available.

Cuttings from normal leaves such as Cape cowslips (*Lachenalia*), snowdrops (*Galanthus*) and snowflakes (*Leucojum*) tend to wilt quickly so keep them turgid by planting as soon as possible. With succulents, such as mother-in-law's tongues (*Sansevieria*), water loss is not so critical.

Leaf cuttings from bulbous plants, such as *Hyacinthus*, that have tender leaves may well rot and die unless they are handled as little as possible, planted carefully and sprayed regularly with fungicide.

Ensure all tools and equipment are absolutely clean before taking leaf cuttings. Fill a container with cuttings compost, press down to within ⅜ in of the container rim, using a presser board. Water the compost thoroughly and allow to drain.

Cut off a fully expanded, undamaged leaf from the parent plant. Lie it face down on a clean sheet of glass and cut with a safety razor blade at right angles to the veins. Make a series of slices 1–1¼ in wide.

Make a shallow trench with a dibber and plant the cutting vertically with its basal cut held firmly in the compost. Place the other cuttings 1 in apart in rows. Label the container clearly. Spray the cuttings with a systemic or copper fungicide as protection against rot and disease.

Place the container in a warm (21°C/70°F), humid environment so the leaves do not dry

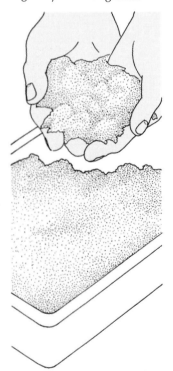

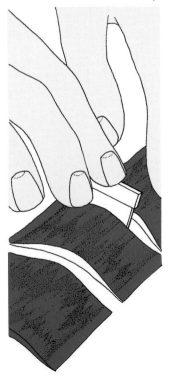

1 Fill a container with cuttings compost. Water well and allow to drain.

2 Cut off a fully expanded, undamaged leaf. Lie face down on some glass.

3 Make a series of 1–1¼ in wide slices at right angles to the veins.

Some monocotyledonous plants responding to leaf cuttings
Galanthus
Heloniopsis
Hyacinthoides
Hyacinthus

Lachenalia – especially
  L. aloides 'Nelsonii'
Leucojum
Sansevieria trifasciata – but not
  S. trifasciata 'Laurentii'
S. zeylanica
Scilla

out and wilt. Ensure the cuttings have adequate light to produce food, although direct sunlight should be avoided otherwise scorching may occur. A polythene tent will provide these conditions quite adequately, although a closed case in a greenhouse will be the most satisfactory provided that it is kept scrupulously clean.

Rewet drying compost by standing the base of the container in a water bath.

The time taken for young plantlets to appear will vary with the different plants. Mother-in-law's tongues will usually regenerate in six to eight weeks during the summer. Cape cowslips, snowdrops and hyacinths that have been propagated in spring just as their leaves mature will regenerate in four to six weeks. Repot the new plants once they are large enough to handle and label them clearly; then harden off.

LOOPING LEAF CUTTINGS

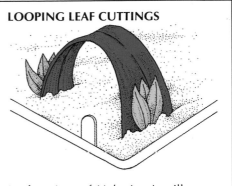

Leaf cuttings of *Heloniopsis* will regenerate from both ends. Reduce the leaf to 1½–2 in by cutting off the top and bottom. Plant the cutting in a loop, with both ends in the compost.

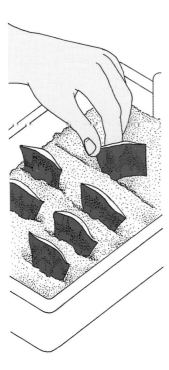

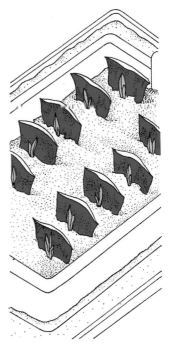

4 Plant the cuttings vertically in trenches 1 in apart. Firm in and label.

5 Spray with a fungicide. Place the container in a warm, humid environment.

6 Rewet the compost if it dries out by standing the container in a water bath.

Foliar embryos

A few plants are capable of developing isolated groups of simple cells in certain areas of their leaves. As a result these cells, or foliar embryos, are capable of developing into new plantlets. Given certain growing conditions some of these plants, such as *Mitella*, have foliar embryos that develop naturally into plantlets. Other plants such as *Cardamine* will respond in this way only if the leaves are separated from the parent plant.

The position on the leaf of these embryos is fixed according to each plant's characteristic and it is not influenced by the way the actual plantlets develop. The plantlets in *Kalanchoe* leaves, for example, arise between the jagged edges of the leaves; in *Mitella* and *Cardamine*, the plantlets appear at the junction of the leaf-stalk and leaf-blade. *Sedum*, however, produces only one plantlet and this is at the base of each sessile leaf.

Collect the plantlets from those plants that produce them naturally. With some of these plants, such as *Kalanchoe*, the plantlets fall off once the roots begin to develop. Plant them in cuttings compost in a labelled seed tray. Repot them separately once they have established properly.

Although other plants such as *Asplenium* and *Cystopteris* develop foliar embryos naturally, it is best to remove the leaf together with the plantlets to allow them slightly longer to become established before separating them from the parent leaf. Place the leaf flat on some cuttings compost in a seed tray. Pin it in position with a light wire staple if it does not sit flat on the compost. Label and leave on a shaded greenhouse bench. Ensure that the parent leaf does not become desiccated. Separate the plantlets and pot on once they have rooted and established. This should take seven to eight weeks.

Some plants, such as *Tiarella*, will be stimulated into developing plantlets from foliar embryos only if their leaves are severed. Cut off a leaf as soon as it has expanded fully. Set it on some cuttings compost in a seed tray. Label; then place in a warm (21°C/70°F), humid, shady environment such as a polythene tent until the plantlets develop and establish (five to seven weeks); then separate and pot on. Succulents, such as *Sedum* and *Kalanchoe*, can be left on an open bench in a greenhouse to develop their plantlets. Certain *Kalanchoe* respond best if the leaves are stimulated in spring to produce plantlets.

Stimulating plantlet development

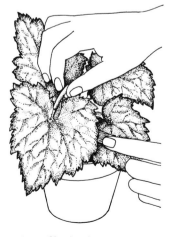

1 Cut off a leaf as soon as it has expanded fully. Fill a container with some cuttings compost.

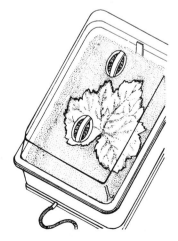

2 Set the leaf on the compost. Label. Place in a warm (21°C/70°F), humid, shaded environment.

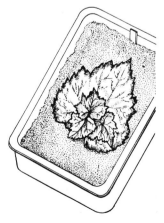

3 Leave until the plantlets develop and establish. Then separate, pot on and label.

Some plants that develop foliar embryos naturally
Asplenium
Cystopteris
Mitella
Tolmiea menziesii
   (pig-a-back plant)

Some plants that require leaf separation for stimulation of plantlet development
Cardamine
Kalanchoe
Sedum
Tiarella

Establishing naturally produced plantlets

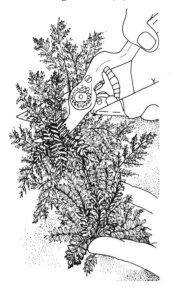
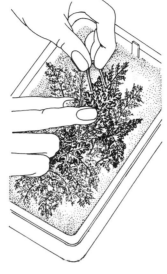
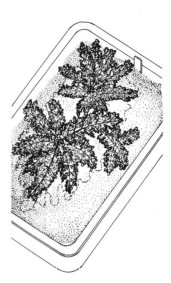

1 Cut off a leaf together with its plantlets. Fill a container with some cuttings compost and firm.

2 Pin the leaf flat on the compost with a wire staple. Label and place on a shaded greenhouse bench.

3 Ensure the leaf does not dry out. Separate the new plantlets and pot on once they are rooted.

WHERE FOLIAR EMBRYOS CAN DEVELOP

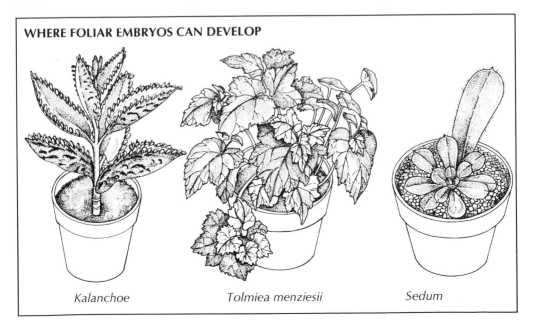

Kalanchoe   *Tolmiea menziesii*   *Sedum*

Grafting/Whip-and-Tongue 1

Grafting is a technique of joining two parts of different plants together in such a way that they will unite and continue their growth as one plant.

One part, called the scion, is usually a stem from the plant to be propagated. This is grafted on to a root system from another plant, which is called the rootstock (also, the stock or understock). All the various techniques of joining plants are called grafting, although, when buds only are joined to the rootstock, it is sometimes called budding.

There are two basic grafting positions: apical grafting, in which the top of the rootstock is removed and is replaced with the scion; and side grafting, in which the scion is grafted on to the side of the rootstock and the rootstock above the graft is not removed until after a union is achieved.

Because time-consuming preparation work is necessary before two plants can be joined, grafting is superficially a less attractive technique than other relatively easy methods of vegetative propagation, such as taking stem

THE ADVANTAGES POSSESSED BY VARIOUS ROOTSTOCKS

Rootstock	Advantage	Plant
East Malling series (EM)	Controls vigour	Apple
Malling Merton series (MM)	Resists woolly aphid and controls vigour	Apple
Colt	Controls vigour	Cherry
East Malling quince	Controls vigour	Pear
Pixie, St Julien A	Controls vigour	Plum
Rhododendron 'Cunningham's White'	Tolerates up to pH7	Rhododendron
Rosa chinensis 'Manetti'	Tolerates soil moisture and salt concentration	Roses (pot-grown)
Rosa laxa	Does not produce suckers, has few thorns and has long hypocotyl	Roses

Whip-and-tongue grafting

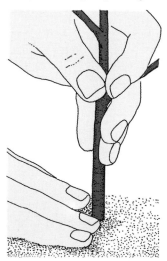

1 Select a plant that will be suitable as rootstock. Plant it outdoors. Label and leave for a growing season.

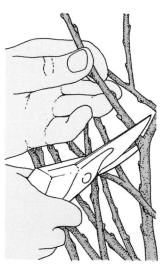

2 Select a plant that is suitable as scion material. Cut off some vigorous hardwood stems for scions.

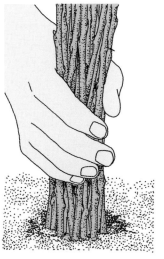

3 Bundle these scions together. Heel them into the ground in a well-drained, cool place. Label.

cuttings or layering. However, some plants such as witch-hazel are not easily propagated by any other vegetative method if selected forms are required, and so they are grafted on to rootstocks.

Perhaps the most useful reason for grafting is to transfer the benefits of a particular rootstock on to another plant. Various fruit-tree rootstocks, for example, have been developed to control both the size and fruiting vigour of other varieties of fruit tree. Other advantages that a rootstock might possess are resistance to pests and diseases; toleration of high soil-moisture levels and salt concentrations; and toleration of high alkalinity levels in the soil. The more rootstock incorporated into a new plant, the more influence the rootstock will have.

Another advantage of grafting is that more than one scion can be joined on to a plant. This is particularly useful with fruit trees as a suitable pollinator variety can be grafted into a tree or bush already grafted with another variety, or it may allow a decorative stem to be grafted on and then be top worked with another variety.

There are a number of problems associated with grafting. The main one is ensuring two plants are compatible. This limitation determines which variety and species of plants can be grafted on to which rootstocks; in general, it is normal to graft varieties on to their own species or very closely related ones.

To graft successfully, it is vitally important to position the various tissues of the stem correctly so that the stem can make a quick and continuous union. The cambium is the actively growing part of the stem that lies just under the bark. This cambium layer on both the scion and the rootstock must be positioned so they are absolutely adjacent to each other, or at least in as much contact as possible.

The successful formation of a graft also depends on making and matching cuts quickly and cleanly: the cut surfaces must be placed in contact with the minimum of delay. Should the surfaces dry out, the tissues will die and so make an effective barrier to the development of a successful union.

The making of a graft union is only partly due to successful carpentry. Much also

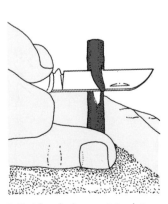

4 Trim the bottom 12–15 in on the rootstock of all branches just before the leaf-buds break.

5 Cut back the rootstock to where the scion is to be grafted. Make a 1½ in sloping cut at the top.

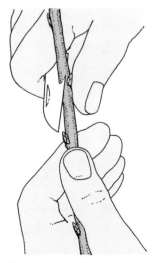

6 Lift the scions from the ground. Make a top cut just above a bud about four buds from the base.

Grafting/Whip-and-tongue 2

depends on providing suitable conditions for the tissues to develop and grow to form a successful union. In effect, this means that water loss must be prevented and warmth must be provided round the grafted parts by carefully covering them until they have joined together.

Traditionally, the grafted parts have been covered and tied together with raffia, but most grafts are now enclosed by ½ in wide clear polythene strip. This has the advantage of completely surrounding and sealing the cut areas, so reducing water loss to a minimum. The graft union on indoor grafted plants is taped with rubberized strip before being placed in a humid atmosphere to develop.

Once the grafted plants have successfully united, the development of the new plant depends on preventing any further competition from the rootstock. Therefore, always remove all subsequent growths from anywhere on the rootstock.

Although it is theoretically possible to graft at almost any time of the year, the best season for most grafting is in the spring.

Shield-budding, however, must normally wait until midsummer when the bark lifts easily from the wood on the rootstock.

Whip-and-tongue grafting

Whip-and-tongue grafting is commonly used to propagate fruit trees, although the technique can be employed successfully for trees and shrubs with tissues that will also readily unite at relatively low temperatures.

Select a plant that will be suitable as rootstock and plant it outdoors. Label it and leave it to establish for one growing season.

In mid-winter, select a plant that is suitable as scion material. From it take some hardwood stems with all their previous season's growth. Bundle these scions together and heel them in 6 in of soil in a well-drained cool position. Firm back the soil and label them. When the scions are eventually grafted in spring, they will be less developed than the growth on the rootstock.

Prepare the rootstock once its sap has started to rise; this is usually just before the leaf-buds break. Trim to make a single stem with no branches.

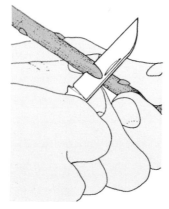

7 Make a 1½ in basal cut at the same angle as the rootstock cut. End it just below the bottom bud.

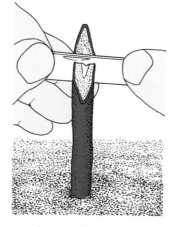

8 Make a shallow, single ½ in slice into the rootstock from one-third down the sloping cut.

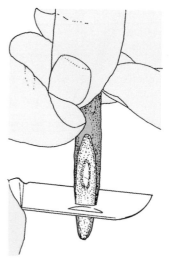

9 Make a shallow, single ½ in slice into the scion from one-third up the scion sloping cut.

Fruit trees (especially apples) are grafted at 9–10 in above soil level to avoid problems such as collar rot. Ornamental plants, on the other hand, are grafted as close to the ground as feasible, so that the unsightly bulges that may occur with certain rootstock/scion combinations are not obvious.

Cut back the rootstock to the appropriate height with a pair of sharp secateurs. Then make a 1½ in sloping cut across the top, using a sharp knife.

Lift the scions from the ground. Select one that has a similar diameter at its base to that of the rootstock top. Take a sharp knife and make a top cut close above a bud about four to five buds from the scion base. Then make a sloping 1½ in cut across the base of the scion, ending it just below a bud; ensure the cut is at the same angle as the rootstock cut.

This is a slice or whip graft. To provide rigidity, add a tongue to the cuts.

The tongue is made from single cuts on both the scion and rootstock. From one-third of the way down the sloping cut on top of the rootstock, make a shallow, single ½ in slice down into the rootstock.

Make the scion tongue by cutting for ½ in from one-third of the way up the scion sloping cut, keeping the knife blade at the same angle as the tongue on the rootstock.

Slip the scion into the rootstock so they interlock. If the rootstock is thicker than the scion, move it to one side until there is good contact between the two cambial layers. Bind with clear polythene grafting tape to hold the join firmly. Seal the top of the scion with tree paint and label.

For a wide range of trees and shrubs, including apple and pear trees, the grafted parts can then be left to unite. Cherry trees, however, should have their scion and grafted area covered with a polythene bag, which is then tied just below the union – the increased temperature hastening development.

Remove the grafting tape and polythene bag as soon as the cut surfaces start callusing, which means the two grafted parts are beginning to join.

Cut off any growth that the rootstock may produce and, if required, reduce the scion shoots to just one to promote a single-stemmed tree or shrub.

10 Slip the scion into the rootstock so they interlock.

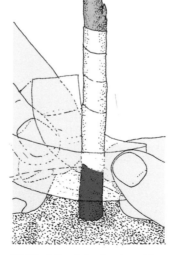

11 Bind the join firmly with clear polythene tape. Dab the top of the scion with tree paint. Label.

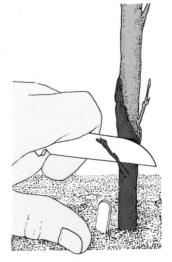

12 Remove the polythene tape once the cut surfaces start callusing. Cut off any rootstock growth.

Apical-wedge

Wedge grafting is perhaps the easiest way to join two plants, and it is used to propagate a wide variety of shrubs and ornamental trees.

In mid-winter, collect some stems with all their previous season's growth from a plant with suitable scion material. Bundle these together and then heel them 6 in deep into the ground. This will prevent the scions drying out, and will keep them cool and retard their development. Label them clearly.

In late winter/early spring, select a suitable rootstock: either a one-year-old seedling or a substantial piece of root that is growing vigorously. Lift the rootstock and wash it.

Grafting is carried out on the top of the root or in the hypocotyl of the seedling. Cut the top of the rootstock horizontally, using a sharp knife. Then make a single 1¼ in cut vertically down the middle of the rootstock.

Lift the scions; choose one that has plump, healthy buds and is of similar diameter to the top of the rootstock. Make a sloping cut just above a bud at the top, using a sharp knife. Then make a horizontal cut about 6 in below. To form a wedge, make a sloping 1½ in cut, starting near a bud and cutting towards the middle of the scion base. Make a similar cut on the opposite side of the scion.

Push the scion gently but firmly down into the rootstock cut. Leave a small portion of the cut scion surface exposed above the rootstock. This will encourage the development of callus tissues, and it is known as the "church window" effect. Bind the joined area tightly with clear polythene grafting tape.

Place the grafted parts in a box of peat and grit and heel them in to just above the graft union. Label and place in a protected environment such as a cold frame or closed case, or on a greenhouse bench. The higher the temperature the faster the union will occur, providing the grafted parts do not dry out.

As the grafted parts unite, the exposed cut surfaces will start callusing. This can be seen in the "church window" just above the cut surface of the rootstock, where the callus tissues interlock and provide rigidity.

When a fairly firm union has formed, cut and remove the polythene tape. Pot up the resulting tree or shrub, or plant it out,

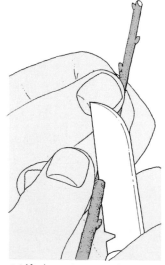

4 Lift the scions. Make a sloping top cut just above a bud and a horizontal one about 6 in below.

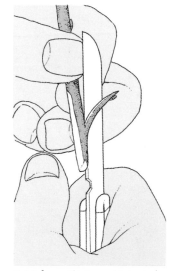

5 Make a 1½ in cut towards the middle of the scion base. Repeat on the other side.

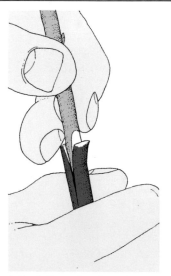

6 Push the scion into the rootstock cut. Leave part of the scion cut surface exposed.

Some plants suitable for apical-wedge grafting
Beech
Cherry, flowering
Cotoneaster, large-growing
Crab apple

Hibiscus
Horse chestnut
Parthenocissus
Rowan
Thorn
Wisteria

depending on its vigour. Write a label in indelible ink and attach it to the new plant. Cut out any competing growths from the rootstock, although these are unlikely to occur with a rootstock that was cut back into the hypocotyl or a root.

1 Collect some stems of the previous season's growth in mid-winter. Heel these scions into the ground. Label.

2 Select a suitable rootstock in late winter/ early spring. Lift and wash it. Cut the top horizontally.

3 Make a single 1¼ in cut vertically down the middle of the rootstock.

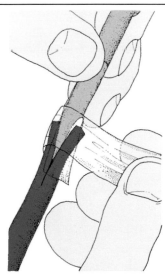

7 Bind the joined area with clear polythene tape.

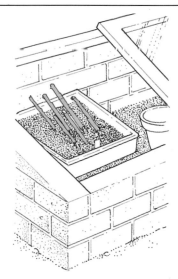

8 Heel the grafted parts into a box filled with peat and grit. Label and place the box in a protected area.

9 Remove the polythene tape once the cut surfaces start callusing. Pot up or plant out.

Side-wedge

The side-wedge graft is the simplest and most effective method of side grafting, and it is widely used to propagate both deciduous and evergreen plants.

The usual season for side-wedge grafting is in late winter/early spring, just before the leaf-bud break. However, with evergreen plants it is feasible whenever a mature flush of growth is available and the rootstock is growing actively. For very sappy plants, that is those plants that "bleed" when cut, it is advisable to graft early in the growing season and dry the rootstocks off before doing so.

Establish a one-year-old labelled seedling in a pot and grow it for a year. Ensure this seedling, which will form the rootstock, is related to and therefore compatible with the plant to be grafted on to it.

About three weeks before grafting, place the potted rootstock in a frost-free area to encourage it to grow. Dry off the rootstock, especially deciduous rootstock, by keeping watering to a bare minimum or indeed by not watering at all. Trim the leaves off the bottom 3–4 in of the rootstock stem.

Select a plant that has suitable scion material and remove some stems with all their previous season's growth. If at all possible, collect these stems with their apical buds intact.

Select a stem, or scion, of comparable thickness to the rootstock. Make two sloping 1½ in cuts on the scion base, opposite each other, so that a wedge shape is produced.

Starting about 2 in above soil level, cut downwards into the rootstock for 1½ in and inwards to about a third of the stem thickness.

Gently bend the rootstock away from this cut so that it opens sufficiently to insert the scion. If the scion is narrower than the rootstock, match up the cambial layers on one side. Release the tension on the rootstock.

Bind with clear polythene tape, overlapping it to seal the entire cut area and label.

Stand the grafted rootstock in a greenhouse, and the two parts should join in about six to eight weeks, depending on the species. This environment is preferable to a warm, humid one, such as a closed case, where the buds would be encouraged to develop and, once the two parts have joined, the grafted rootstock would need hardening off, which can be extremely difficult to do satisfactorily.

Remove the polythene tape and cut back half the rootstock above the grafted area as soon as the two parts have joined. Two weeks

4 Make two sloping 1½ in cuts, opposite each other, on the scion base.

5 Start about 2 in above the ground and cut downwards and slightly inwards for 1½ in into the rootstock.

6 Bend the rootstock gently away from this cut. Hold it there and insert the scion. Then release it.

Some plants suitable for side-wedge grafting
Birch
Horse chestnut

Hazel
Japanese maple
Rhododendron
Witch-hazel

later, cut back the remaining rootstock above the grafted area so that the scion becomes the leader shoot of the plant. Dab the cut surface with a suitable tree paint if the final cut surface is extensive.

Watering is probably the most difficult aspect of graft management. Water little and often rather than in large doses at infrequent intervals, so that the plant is kept on the dry side in a humid environment.

1 Establish a one-year-old seedling in a pot and label it. Grow it for one year.

2 Dry off this rootstock in a frost-free area just before bud-break. Trim the leaves off the bottom 3–4 in.

3 Select a plant that has suitable scion material. Cut off some stems with all their previous season's growth.

7 Bind the entire cut with clear polythene tape and label. Stand the grafted rootstock in a greenhouse.

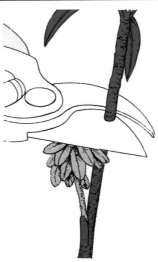

8 Remove the tape and head back half the rootstock once the two parts have joined.

9 Cut back the remaining rootstock above the grafted union two weeks later.

177

Side-veneer

Traditionally, side veneer grafts have been used to propagate conifers, but they are now often carried out on any plant suitable for side-grafting. The join resulting from this method of grafting is slightly more upright than one from a side-wedge graft (see page 176).

Conifers, when cut, exude resin and, although this can be reduced by drying off the rootstocks before grafting, the resin will still adhere to a knife blade and so impair its efficiency and cutting edge. Therefore, keep the knife blade clean by wiping it regularly with some absorbent material soaked in an organic solvent such as acetone. It is important to wipe the solvent off the blade with a clean cloth otherwise it could pollute the tissues of any plant through which the knife blade subsequently cuts.

Rootstocks for side-veneer grafts are almost invariably pot-grown, as the fine, fibrous roots of the conifer are prone to drying out.

Graft conifers either in late winter/early spring or during late summer, although the latter period causes problems as the tree has to kept alive over the winter and then needs to be hardened off in spring.

In late winter/early spring, dry off the pot-grown conifer that will form the rootstock by watering less than usual. Because it is evergreen, this drying off should not be as drastic as for deciduous plants.

After three weeks, once the rootstock shows signs of growth, grafting is carried out.

Select a conifer that has suitable scion material and cut off a leader shoot of one season's growth. The length of the shoot, or scion, will very much depend on the vigour of growth, but it must include some mature brown-barked wood. Strip off the leaves from the bottom third of the scion. Make a 1¼ in sloping cut down to the scion base, using a sharp knife. Then turn the scion over and make a very small wedge cut.

Trim any leaves off the bottom 6 in of the rootstock stem. About 2½–3 in above soil level, make a 1¼ in sloping cut downwards and inwards to give a shallow cut about ⅛ in deep at the bottom. Then make a slightly downward and inward nick lower down the rootstock to join the first cut. Remove the sliver of wood from the rootstock. Place the rootstock and scion together. If their sizes match they will be easier to manipulate. If the scion is narrower than the rootstock, set to one side to match up the cambial layers. The interlocking bases of the two grafted parts will provide some rigidity, but they should also be tied together firmly.

It is possible to use a rubberized strip as a tie, but clear polythene tape is just as effective and has the added advantage of preventing water loss from around the cuts.

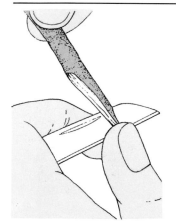

4 Make a 1¼ in sloping cut down to the scion base. Make a very small wedge cut on the other side.

5 Trim any leaves off the bottom 6 in of the rootstock. Make a 1¼ in shallow cut at the base.

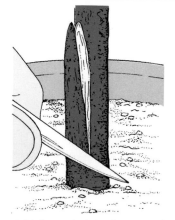

6 Make a slightly downward and inward nick lower down the rootstock to join the first cut.

Label the grafted rootstock and keep it warm and humid to encourage the plant to grow and the grafted parts to join up. A closed case with or without heat from below is a suitable environment.

Water the pot sparingly, keeping the plant on the dry side until callusing occurs.

Conifers, if grown close together in pots under warm, humid conditions, are subject to various leaf diseases and rots, and to red spider mite infestations, so spray regularly with fungicides and pesticides.

Take the grafted rootstock out of the closed case and harden off on the greenhouse bench, once the parts have united. This should be within six to ten weeks. Remove the polythene tape and reduce the top of the rootstock by one-third immediately. After a fortnight, remove a further third. After mid-summer, head back any remaining stock.

1 Select a pot grown conifer that will be suitable as rootstock. Dry it off in late winter.

2 Select a conifer that has good scion material. Cut off a strong shoot that has some mature wood at its base.

3 Strip off the leaves from the bottom third of the scion.

7 Remove the sliver of wood. Place the rootstock and scion together and tie with polythene tape. Label.

8 Place in a warm, humid area while the two parts unite. Spray regularly with fungicides and pesticides.

9 Harden off once the parts have united. Remove the tape and gradually head back the rootstock.

Shield-budding

Shield-budding, or T-budding, is a traditional way to propagate rosaceous plants by grafting. A bud from the plant to be propagated is placed behind the bark of the rootstock so that the back of the bud and the exposed surface of the rootstock wood are in contact. However, this technique can only be carried out when the bark of the rootstock lifts easily; this is normally between May and August.

Select a suitable rootstock, either a seedling or a one-year-old layer. Plant it in some open ground during the winter. Label it and allow to establish.

As soon as the bark lifts easily from the wood underneath, prepare the rootstock for shield-budding. Insert a rose bud into the hypocotyl of a seedling rose (see page 182) and a tree bud into a stem of a compatible rootstock at the required height.

Cut off any leaves and branches on the bottom 12 in of the rootstock. Make a T-shaped incision through the bark by cutting a horizontal slit and then a vertical downward incision sufficiently large to take a suitable bud. The taut bark will begin to spring away from the wood underneath. Loosen the two flaps slightly to receive the bud.

Select a plant that has suitable budding material. Cut off a stem with all the current year's growth and with plump, healthy buds.

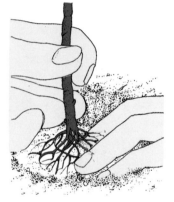

1 Plant a rootstock in the open ground in winter. Label. Allow to establish.

2 Trim the bottom 12 in of the rootstock of all leaves and branches in summer.

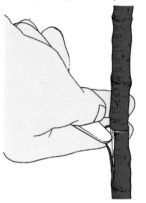

3 Make a T-shaped incision through the bark. Loosen the two flaps of bark.

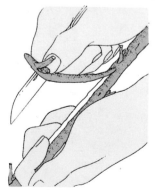

7 Cut shallowly underneath the bud. Lift it off once the knife is past.

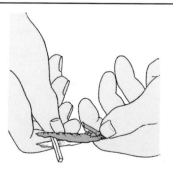

8 Bend the bark outwards and flick out any wood underneath the bark.

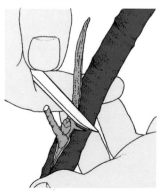

9 Slip the bud into the T-cut on the rootstock. Trim the tail neatly.

Remove the leaves, but leave about ½ in of each leaf-stalk attached to the stem.

When shield-budding, always select buds from the middle of a stem where the buds are mature. Do not take them from the bottom of the stem because they may be latent, nor from the top where they will be immature.

Cut shallowly into the stem, or bud-stick, about ¼ in below a mature bud; then cut shallowly underneath the bud; when past it, lift off the bud together with a tail of bark. Ensure the cut is deep enough to avoid damaging the "eye" of the bud.

Remove any wood from under the bark by bending the bark outwards and flicking the wood out. If the bud-trace comes out with the wood, the bud is not mature and should be discarded.

Using the leaf-stalk as a handle, slip the bud into the T-cut on the rootstock, and trim off the tail flush with the horizontal cut.

Tie the budded rootstock with clear polythene tape, leaving the bud and leaf-stalk exposed, and label it.

After three to four weeks the bud will have united with the rootstock and the tape can be removed.

In late winter/early spring, cut back the top of the rootstock to just above the bud, which will then grow out during the following spring.

4 Select a plant that has suitable budding material. Cut off a vigorous stem.

5 Remove the leaves but retain ½ in of each leaf-stalk on the stem.

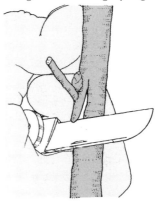

6 Cut shallowly into the stem about ¼ in below a mature bud.

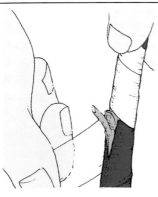

10 Tie with polythene tape, leaving the bud and leaf-stalk exposed. Label clearly.

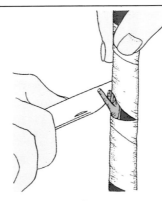

11 Remove the tape once the bud has united with the rootstock.

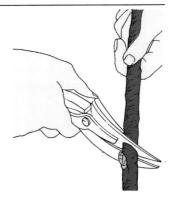

12 Cut back the top of the rootstock in late winter/early spring.

Rose budding

Hybrid Tea and Floribunda roses should be propagated by inserting buds into seedling rootstocks rather than by taking stem cuttings. The advantage of rose budding is that the seedling rootstocks boost the vigour of the weaker varieties, which on their own roots may grow only 9 in or so, while reducing the potential of the more vigorous ones, which might otherwise reach 14–16 ft tall.

The selection of a suitable rootstock is one of the most important decisions when propagating roses. Although it is possible to collect a seedling of the wild rose from the hedgerow and use this as rootstock, this not only denudes the natural flora of the countryside but in fact is not really desirable as rootstock.

A rose rootstock should not produce suckers once the bush is eventually planted in its permanent position. It should ideally be of a variety that grows few thorns, and the actual process of budding will be easier if the neck of the rootstock is relatively long. All these advantages are found in *Rosa laxa*.

In winter, plant a seedling rootstock of 5–8mm grade up to its neck in the ground. Space further plants about 9 in apart in a row, and earth them all up to cover their necks; then label them.

Budding is carried out once the bark of the rootstock lifts easily, usually after midsummer.

Pull away the earth from the rootstock neck. Make a T-shaped cut in the neck. Loosen the two flaps of bark a little.

Select a plant that has suitable budding material. From it choose a stem in which the flowers have just "blown", that is, gone over, and remove this stem with all its current season's growth. At this stage virtually all the buds on the stem wil be suitable for propagation. Remove the leaves from the stem.

Hold a sharp knife just below a bud. Cut shallowly towards the tip of the stem round the bud. Lift off the bud together with its tail.

Insert the bud, tail upwards, between the flaps of bark on the rootstock. Trim the tail flush with the horizontal cut. Cover the bud with a rubber budding patch and pin it in position. This patch, which is made of non-vulcanized rubber, will perish at about the same rate as the bud unites with the rootstock – in about four to five weeks. This means that the rubber patch does not need cutting as it will not constrict the rapidly expanding neck of the plant.

The bud will develop before the end of the season or during the following season.

In late winter, cut off the top of the rootstock, just above the bud. This prevents possible stem suckers. The following autumn, replant the bush in its final place.

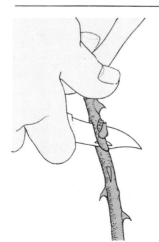

4 Cut shallowly towards the tip of the stem round a bud. Lift off the bud.

5 Insert the bud, tail upwards, between the flaps of bark on the rootstock.

6 Trim the tail flush with the horizontal cut.

Standard roses

Standard roses are propagated on to root-stocks of *Rosa rugosa*, which has single stems usually 6–8 ft tall. Plant in winter in a row and support on wire attached to posts. Budding takes place during the following summer when the bark lifts. Two or sometimes three buds in a close spiral are worked on the root-stock in order to produce a more even and regular head. The height at which the leaf-buds are inserted will depend on the length of stem required. A standard rose is usually budded at about 3½ ft; a half-standard at about 2¾ ft.

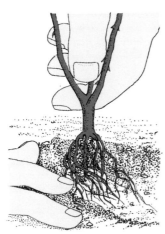

1 Plant a seedling rootstock of 5–8 mm grade up to its neck in the open ground in winter. Earth up and label it clearly.

2 Pull away the earth from the rootstock neck after midsummer. Make a T-shaped cut at the neck. Loosen the flaps of bark.

3 Select a plant that has suitable budding material. Cut off a stem in which the flowers have just "blown". Remove all the leaves.

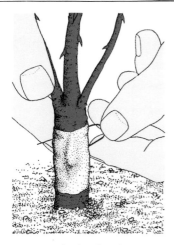

7 Cover the bud with a rubber budding patch. Pin in position; then label.

8 Cut off the top of the rootstock just above the bud in later winter.

9 Lift and plant out the new rose bush the following autumn.

Chip-budding

Chip-budding is perhaps the easiest way to bud a plant as it involves relatively few actions and, more importantly, it provides greater cambial contact between rootstock and scion than does shield-budding.

A "chip" of bark and wood is removed from the rootstock and replaced with a "chip" of similar dimensions carrying a bud from the plant to be propagated.

Chip-budding can be used to propagate any rosaceous plant, provided that the wood is sufficiently hard and mature. For the technique to be successful, it is necessary to prevent any water loss by sealing the cut edges extremely well.

Chip-budding can be carried out at any time of the year provided that well-matured buds are available and temperatures are high enough (at least 10°C/50°F) to produce a union quickly.

Although this technique could be used on pot-grown rootstocks such as *Magnolia grandiflora*, it is usually carried out on rootstocks that have been planted in winter in the open ground.

In midsummer, remove all branches and leaves from the bottom 12–15 in of the rootstock stem.

Select a plant that is suitable for budding. Cut off some vegetative shoots with all their current year's growth and with well-matured buds, at least towards their base. Discard the softer top growth and carefully remove all the leaves flush with the stem.

Make a ¼ in cut down into the rootstock stem, at an angle of about 45 degrees. Start a top cut 1¼ in above the lower cut and angle it down to join the basal cut; remove the chip.

Select a stem, or bud-stick, that has a similar diameter to the rootstock so that it is easy to match the cuts.

Make exactly similar cuts on the bud-stick as on the rootstock, ensuring that a bud is included midway down the chip.

Tuck the bud-chip into position on the rootstock and tie with polythene tape. Wind the tape round the rootstock, overlapping so that it seals the chip completely. Then label.

After three to four weeks the bud will have united with the rootstock and the tape can be removed, so allowing the bud to swell.

In winter, cut the rootstock right back, close above the bud but without damaging it. The bud will grow out the following season. If budding is done early, the bud may grow out before the end of the season.

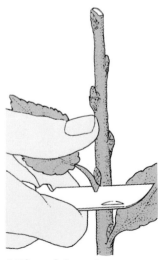

4 Discard the softer top growth on each bud-stick. Remove all the leaves flush with the stem.

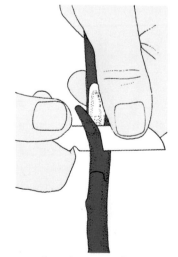

5 Make a ¼ in cut down into the rootstock stem. Slice down to it from 1¼ in up the stem. Remove the chip.

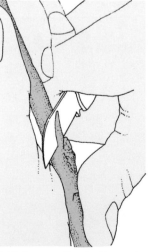

6 Make exactly similar cuts on the bud-stick, ensuring that a bud is included midway down the chip.

1 Plant a suitable rootstock in the open ground in winter. Label it clearly.

2 Remove all branches and leaves from the bottom 12–15 in of the rootstock in midsummer.

3 Select a plant that is suitable for budding. Cut off some strong vegetative shoots with mature buds.

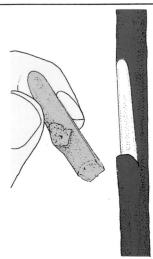

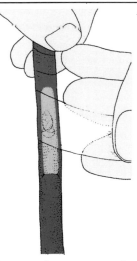

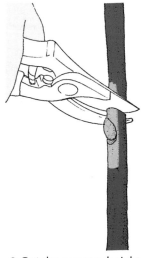

7 Tuck the bud-chip into position on the rootstock. Cover the chip with clear polythene tape; then label.

8 Remove the tape once the bud has united with the rootstock.

9 Cut the rootstock right back, close above the bud, in winter. The bud will grow out the next season.

Index/Acknowledgements

In a book of this type it is not possible or appropriate to mention every plant by name in the text. Not every plant listed below, therefore, has a specific reference on the page(s) given, but the techniques and information described on the given pages are those suitable for the plant concerned.

Abelia 135
Abies
    species 66–8
    varieties 178–9
Abutilon
    indoor 130–1
    outdoor 126–9
Acacia 62–3, 66–9, 74–9
Acaena 98–101
Acantholimon 98–101, 152–3
Acanthus 50–1, 74–9
Acer
    species 54–5, 66–9
    Japanese maples 176–7
    Tree maple varieties 116–17, 180–1
Achillea
    alpine 130–1
    herbaceous perennial 98–101
Achimenes 84–5
Acorus 98–101
Actinidia 126–9, 132–3
Adenophora 46–7
Adiantum 52–3, 98–9
Adonis
    alpine 46–7
    herbaceous perennial 50–1
Aechmea 102–3
Aesculus
    species 66–9, 74–9
    varieties and hybrids 176–7, 180–1
Aethionema 130–1
Agapanthus 98–101
Agave 102–3
Ageratum 48–9
Aichryson 130–1
Ailanthus 66–9, 74–9
Ajuga
    alpine 98–101, 102–3
    herbaceous perennial 102–3
Akebia 134–7
Alchemilla 98–101
Allemanda 130–1
Allium
    alpine 46–7, 98–101
    bulbous 46–7, 90–1, 98–9
    herbaceous perennial 90–1, 98–101
Almond 184–5
Alnus 66–9
Aloe 102–3
Alyssum
    alpine 46–7, 130–1
    annual bedding plant 48–9
    herbaceous perennial 98–101, 130–1
Amaranthus 48–9
Amaryllis 90–1, 98–101

Amelanchier 64–5, 66–9, 180–1
Ampelopsis, 130–1, 144–5
Anacyclus 98–101
Anagallis 98–101
Ananas 102–3
Anaphalis 98–101
Anchusa 74–9
Andromeda 118–19, 154–5
Androsace 46–7, 102–3, 132–3
Anemone
    alpine 46–7, 98–101
    bedding plant 48–9, 80–1
    herbaceous perennial 50–1, 74–9, 98–101
    tuberous 46–7, 86–7
Antennaria 98–101
Anthemis
    alpine 130–1
    herbaceous perennial 50–1, 98–101
Antholyza 98–101
Anthurium 98–101
Anthyllis 132–3
Antirrhinum 48–9
Aphelandra 122–3, 130–1
Apple 114–15, 170–3, 180–1, 184–5
Apricot 114–15, 170–3, 180–1, 184–5
Aquilegia
    alpine 46–7
    herbaceous perennial 50–1, 98–101
Arabis
    alpine 46–7, 98–101
    bedding plant 48–9, 130–1
Aralia
    indoor 48–9, 74–9
    outdoor 74–9
Araucaria 66–9
Arbutus 66–9, 134–7
Arctotis 48–9, 98–101, 130–1
Ardisia 48–9, 74–9
Arenaria 98–101
Arisaema
    alpine 46–7, 98–101
    tuberous 86–7, 98–101
Aristolochia 130–1
Armeria
    alpine 98–101, 130–1
    herbaceous perennial 130–1
Arnebia
    alpine 74–9, 98–101
    herbaceous perennial 74–9
Aronia 98–101
Artemisia 150–1
Arum 86–7, 98–101
Aruncus 98–101
Arundinaria 98–101
Asparagus 48–9, 86–7
Asperula 98–101
Asphodeline 50–1, 98–101
Asphodelus 50–1
Aspidistra 98–101
Asplenium 52–3, 98–101, 168–9
Aster
    alpine 98–101
    annual bedding plant 48–9
    herbaceous perennial 98–101

Astilbe 98–101
Astragalus 132–3
Astrantia 98–101
Aubrieta 98–101
Aucuba 134–7
Avocado 70–1
Azalea
    deciduous 108–9, 126–9
    evergreen 132–3, 134–7
Azara 130–1

Ballota 150–1
Baptisia 98–101
Begonia
    fibrous-rooted 48–9
    rex 86–7, 162–3, 164–5
    species 84–5, 158–9
    tuberous-rooted 48–9, 80–1
Bellis 46–7, 98–101
Beloperone 130–1
Berberis 124–5, 132–3, 134–7
Bergenia 98–101
Betula
    species 66–9
    varieties 126–9, 176–7
Bignonia 74–9
Billbergia 98–101
Blackberry 112–13, 122–3, 130–1
Black currant 138–41
Bletilla 98–101
Borago 50–1, 130–1
Bougainvillea 130–1
Boysenberry 112–13, 122–3, 130–1
Brodiaea 90–1, 98–101
Bromelia 102–3
Browallia 48–9
Brunfelsia 126–9
Brunnera 74–9
Bryophyllum 168–9
Buddleja 130–1, 142–3
Buphthalmum 130–1
Buxus 134–7

Caladium 84–5
Calamintha 98–101
Calandrinia
    alpine 46–7
    bedding plant 48–9
Calathea 98–101, 130–1
Calceolaria
    alpine 46–7, 98–101
    half-hardy 48–9
Calendula 48–9
Callicarpa 132–3
Callistemon 40–1, 48–9
Callistephus 48–9
Calluna 118–19, 154–5
Calocedrus 146–9
Calochortus 46–7
Caltha
    alpine 46–7, 98–101
    herbaceous perennial 98–101
Camellia 122–3, 134–7
Campanula
    alpine 46–7, 98–101

herbaceous perennial 50–1, 98–101
Campsis 74–9
Canna 48–9, 62–3, 86–7
Capsicum 48–9
Caragana 66–9, 174–5
Cardamine 46–7, 168–9
Cardiocrinum 46–7, 90–1
Carduncellus 46–7, 74–9
Carlina 46–7, 74–9
Carpenteria 130–1
Carpinus
    species 64–5, 66–9
    varieties 176–7
Carya 66–9
Caryopteris 130–1, 132–3
Cassiope 118–19, 154–5
Castanea
    species 66–9
    varieties 176–7, 180–1, 184–5
Catalpa
    species 66–9
    varieties 74–9, 176–7
Ceanothus
    deciduous 130–1
    evergreen 134–7
Cedrus
    species 66–9
    varieties 178–9
Celastrus 130–1, 138–41
Celmisia 46–7
Celosia 48–9
Centaurea 98–101
Cephalotaxus 66–9, 178–9
Cerastium 46–7
Ceratostigma 130–1
Cercidiphyllum 106–9
Cercis 66–9, 106–9
Cestrum 130–1
Chaenomeles 74–9, 114–15
Chamaecyparis
    species 66–9
    varieties 146–9
Cheiranthus
    alpine 132–3
    biennial bedding plant 48–9
Cherry 170–3, 184–5
Chimonanthus 108–9, 126–9
Chionodoxa 90–1, 98–101
Chlorophytum 102–3
Choisya 134–7
Chrysanthemum
    alpine 98–9
    bedding plant 48–9, 98–101, 130–1
    herbaceous perennial 98–101, 130–1
Cimicifuga 98–101, 130–1
Cineraria 48–9
Cissus 130–1
Cistus 130–1, 132–3
Citrus 70–1, 110–11
Clarkia 48–9
Claytonia 46–7
Clematis
    herbaceous perennial 98–101, 130–1
    species 66–9
    varieties 122–3, 130–1, 174–5

Clerodendrum
    indoor 130–1
    outdoor 74–9
Clethra 130–1
Clianthus 48–9, 62–3
Clivia 98–101
Cobaea 48–9
Codiaeum 130–1
Codonopsis 46–7
Coffea 70–1
Colchicum 46–7, 90–1
Coleus 48–9, 130–1
Columnea 130–1
Colutea 66–9
Comptonia 74–9
Conifers 146–9, 178–9
Convallaria 86–7, 98–101
Convolvulus 132–3
Cordyline 98–101
Coreopsis 50–1, 98–101
Cornus
    coloured-bark 116–17, 132–3, 138–41
    species 64–5, 66–9
Coronilla
    alpine 132–3
    shrub 130–1
Cortaderia 98–101
Corydalis 46–7
Corylopsis 106–9, 126–9
Corylus
    species 64–5, 66–9
    varieties 108–9, 176–7
Cotinus 108–9, 116–17
Cotoneaster
    large-leaved 138–41, 174–5, 180–1
    small-leaved 124–5, 132–3
Cotula 98–9
Crambe 50–1, 130–1
Crassula 102–3, 130–1
Crataegus
    species 62–3, 64–5, 66–9
    varieties 180–1, 182–3
Crepis 74–9
Crinodendron 130–1, 132–3
Crocosmia 88–9, 98–101
Crocus 46–7, 88–9
Cryptanthus 102–3
Cryptomeria 146–9
Cunninghamia 66–9
Cuphea 130–1
× Cupressocyparis 146–9
Cupressus 66–9, 146–9
Cyananthus 132–3
Cyclamen
    alpine 46–7, 80–1
    bedding plant 48–9
    tuberous 46–7, 80–1
Cyperus
    indoor 98–101
    outdoor 48–9, 98–101
Cystopteris 52–3, 168–9
Cytisus
    alpine 132–3
    species 62–3, 66–9
    varieties 132–3

Daboecia 118–19, 154–5
Dahlia
    bedding plant 48–9, 130–1
    herbaceous plant 46–7, 80–1, 130–1
Daphne
    alpine 46–7, 134–7, 176–7
    species 62–3, 64–5, 66–9
    varieties 132–3, 134–7
Date palm 70–1
Datura 130–1
Davallia 86–7
Davidia 64–5, 66–9
Delphinium 50–1, 98–101, 130–1
Dendromecon 74–9
Deutzia 132–3, 142–3
Dianthus
    alpine 132–3
    herbaceous perennial 50–1, 98–101, 130–1
Dicentra
    alpine 98–101
    herbaceous perennial 98–101
Dictamnus 150–1
Dieffenbachia 98–101
Diervilla 132–3, 142–3
Digitalis 50–1
Dionysia 46–7
Dioscorea 84–5
Dipladenia 130–1
Disanthus 106–9
Distylium 106–9, 134–7
Dizygotheca 48–9
Dodecatheon 46–7, 98–101
Doronicum 98–101
Dorotheanthus 49
Douglasia 98–101
Draba 46–7
Dracaena 130–1, 138–41
Dryas 132–3

Eccremocarpus 66–9, 74–9
Echeveria 102–3
Echinops 50–1
Edraianthus 46–7
Elaeagnus 134–7
Embothrium 74–9
Endymion 46–7, 90–1, 98–101, 166–7
Enkianthus 132–3
Epilobium 46–7
Epimedium 98–101
Episcia 130–1
Eranthis 46–7, 86–7
Erica
    indoor 154–5
    outdoor 118–19, 154–5
Erigeron
    alpine 98–101, 130–1
    herbaceous perennial 98–101
Erinus 46–7
Eriogonum 46–7, 130–1
Eritrichium 46–7
Erodium 46–7, 74–9
Eryngium
    alpine 46–7, 74–9
    herbaceous perennial 74–9

Erysimum 46–7, 130–1
Erythronium 46–7, 80–1
Escallonia 134–7
Eucalyptus
   bedding plant 48–9
   tree 40–1
Eucharis 90–1
Eucomis 90–1
Eucryphia 132–3
Euonymus
   dedicuous 66–9, 130–1
   evergreen 134–7
Eupatorium 98–101
Euphorbia
   indoor 130–1
   outdoor 98–101
Euryops 132–3
Exacum 48–9
Exochorda 130–1, 138–41

Fagus
   species 66–9
   varieties 176–7
× *Fatshedera* 122–3, 130–1
Ficus
   indoor 110–11, 122–3, 130–1
   outdoor 122–3, 132–3
Fig 108–9, 176–7
Filbert 106–9
Filipendula 98–101
Fittonia 130–1
Forsythia 130–1, 132–3, 142–3
Fothergilla 108–9
Frankenia 98–101, 132–3
Fraxinus
   species 64–5, 66–9
   varieties 180–1, 184–5
Freesia 46–7, 88–9
Fremontodendron 132–3
Fritillaria 46–7, 94–5
Fuchsia 130–1

Gaillardia 98–101
Galanthus 46–7, 90–1, 96–7, 166–7
Galtonia 46–7, 90–1
Gardenia 130–1
Garrya 134–7
Gaultheria
   alpine 154–5
   shrub 118–19, 154–5
Gazania 48–9, 98–101, 130–1
Genista
   alpine 132–3
   shrub
      species 66–9
      varieties 134–7
Gentiana
   alpine 46–7, 98–101, 130–1
   herbaceous perennial 50–1, 98–9
Geranium
   alpine 46–7, 74–9, 98–101
   herbaceous perennial 74–9, 98–101, 130–1
Gerbera 98–101
Gesneria 160–1

Geum 98–101, 102–3
Ginkgo
   species 66–9
   varieties 138–9, 146–9
Gladiolus 88–9
Gleditsia 62–3, 64–5, 66–9, 176–7
Globularia 98–101
Gloxinia 158–9
Gooseberry 130–1, 142–3
Grape vine 144–5
Grevillea 48–9
Griselinia 134–7
Guzmania 102–3
Gymnocladus 62–3, 66–9, 74–9
Gynura 130–1
Gypsophila
   alpine 98–101
   herbaceous perennial 46–7, 130–1

Haberlea 46–7, 122–3
Habranthus 90–1
Haemanthus 90–1
Halesia 64–5, 66–9, 106–9
Halimium 130–1
Hamamelis
   species 62–3, 64–5, 66–9
   varieties and hybrids 106–9, 110–11, 176–7
Hazel nut 106–9
Hebe
   alpine 134–7
   shrub 120–1, 134–7
Hedera
   indoor 122–3, 132–3
   outdoor 122–3, 134–7
Helenium 98–101
Helianthemum
   alpine 132–3
   shrub 130–1
Helianthus 50–1, 98–101
Heliopsis 130–1
Heliotrope 48–9, 130–1
Helleborus
   alpine 46–7, 98–101
   herbaceous perennial 50–1
Heloniopsis 166–7
Helxine 98–101
Hemerocallis 98–101
Hepatica 46–7, 98–101
Heuchera 98–101
Hibiscus
   bedding plant 130–1
   shrub or tree 66–9, 174–5
Hieracium 46–7
Hippeastrum 78–9
Hippocrepis 130–1
Hippophae 62–3, 66–9
Hoheria 108–9, 116–17
Holodiscus 138–41
Hop 122–3, 130–1
Hosta 98–101
Houstonia 98–101
Hoya 122–3, 134–7
Humulus 122–3, 130–1
Hyacinthus 90–1, 96–7, 166–7

Hydrangea
   indoor 130–1
   outdoor 106–9, 130–1, 132–3
Hymenocallis 90–1
Hypericum
   alpine 130–1
   herbaceous perennial 98–101
   shrub 132–3
Hypocyrta 130–1
Hypsela 98–101

Iberis 130–1
Ilex 134–7
Impatiens 48–9, 130–1
Incarvillea 50–1
Inula 98–101
Ipheion 46–7, 90–1
Ipomoea 48–9
Iresine 130–1
Iris
   alpine 130–1
   bulbous 90–1
   herbaceous perennial 98–101
   rhizomatous 86–7
Isolepis 98–101
Itea 126–9
Ixia 88–9
Ixiolirion 88–9

Jacaranda 48–9
Jancaea 158–9
Japanese wineberry 74–9, 112–13, 122–3, 130–1
Jasione 98–101
Jasminum
   alpine 134–7
   climber 130–1
   houseplant 130–1
   shrub
      deciduous 130–1
      evergreen 134–7
Jeffersonia 98–101
Juglans 66–9
Juniperus
   species 66–9
   varieties
      low-growing 146–9
      tall 178–9

Kalanchoe 48–9, 130–1, 168–9
Kalmia 118–19, 152–3
Kentia 40–1
Kerria 142–3
Kniphofia 98–101
Kochia 48–9
Koelreuteria 66–9, 74–9
Kolkwitzia 130–1, 132–3

Laburnum
   species 64–5, 66–9
   varieties 180–1, 184–5
Lachenalia 90–1, 96–7, 166–7
Lactuca 46–7, 74–9
Lagerstroemia 48–9
Lamium 98–101, 130–1

Lantana 130–1
Lapeirousia 46–7
Larix
  species 66–9
  varieties 178–9
Lathyrus 48–9, 62–3
Laurus 134–7
Lavandula 150–1
Lavatera 98–101, 130–1
Ledum 118–19, 134–7
Leiophyllum 154–5
Leontopodium 46–7
Leptospermum 132–3
Leucogenes 130–1
Leucojum 90–1, 96–7, 166–7
Leucothoë 118–19, 134–5
Lewisia 46–7, 98–101
Leycesteria 142–3
Liatris 98–101
Ligularia 50–1, 98–101
Ligustrum
  deciduous 142–3
  evergreen 134–7
Lilium
  alpine 46–7, 90–1, 92–3
  bulbous 46–7, 90–1, 92–3, 94–5
Limonium 74–9
Linaria
  alpine 46–7, 130–1
  herbaceous perennial 98–101
Linum
  alpine 46–7, 130–1
  herbaceous perennial 98–101
Lippia 130–1
Liquidambar 62–3, 66–9
Liriodendron 62–3, 66–9
Liriope 50–1
Lithocarpus 66–9
Lithospermum 132–3
Lobelia
  bedding plant
    annual 48–9
    perennial 98–101
  herbaceous perennial 98–101
Lobularia 49
Loganberry 112–13, 122–3, 130–1
Lonicera
  deciduous 138–41
  evergreen 134–7
Lupinus
  herbaceous perennial 50–1, 98–101
  shrub 130–1
Lychnis 98–101
Lyonia 118–19, 134–7
Lysimachia 98–101
Lythrum 98–101

Maclura 66–9
Magnolia
  species 62–3, 64–5, 66–9, 110–11
  varieties
    deciduous 110–11, 126–9
    evergreen 110–111, 134–7
Mahonia 122–3, 134–7
Malus

  rootstock 114–15
  varieties 170–3, 174–5, 180–1, 184–5
Malva 130–1
Maranta 130–1
Matthiola 48–9
Mazus 98–101
Meconopsis
  alpine 46–7
  herbaceous perennial 50–1
Medlar 170–3, 180–1, 184–5
Melon 48–9
Merendera 46–7, 88–9
Mertensia
  alpine 98–101
  herbaceous perennial 50–1, 98–101
Mesembryanthemum 48–9
Mespilus 170–3, 180–1, 184–5
Metasequoia 138–41
Metrosideros 126–9, 134–7
Mimosa 48–9
Mimulus
  alpine 46–7, 130–1
  herbaceous perennial 98–101
Mirabilis 48–9
Mitella 168–9
Moltkia 130–1
Moluccella 48–9
Monarda 130–I
Monstera 48–9, 122–3, 132–3
Montbretia 88–9, 98–101
Moraea 90–1, 98–101
Morisia 74–9
Morus 66–9
Muehlenbeckia 134–7
Mulberry, black 66–9, 130–1
Muscari 46–7, 90–1, 96–7, 98–101
Myoporum 134–7
Myosotis 98–101
Myrica 134–7
Myrtus 134–7

Nandina 62–3, 64–5, 66–9
Narcissus 46–7, 90–1, 96–7
Neanthe 48–9
Nectarine 180–1, 184–5
Neillia 138–41
Neoregelia 102–3
Nepeta
  alpine 130–1
  herbaceous perennial 98–101
Nephrolepis 52–3, 98–101
Nephthytis 98–101
Nerine 46–7, 90–1, 96–7
Nerium 130–1, 132–3
Nicotiana 48–9
Nolina 48–9
Nomocharis 46–7, 90–1, 94–5
Nothofagus 66–9
Notholirion 46–7, 90–1, 94–5
Nymphaea 84–5
Nyssa 66–9, 126–9

Oenothera
  alpine 46–7, 98–101
  herbaceous perennial 50–1, 98–101

Olearia 134–7
Omphalodes
  alpine 98–101
  herbaceous perennial 50–1
Omphalogramma 46–7
Ononis 46–7
Oplismenus 130–1
Origanum
  alpine 130–1
  herbaceous perennial 98–101, 130–1
Ornithogalum 46–7, 90–1
Osmanthus 134–7
× *Osmarea* 134–7
Ostrya 64–5, 66–9
Ourisia 98–101
Oxalis
  alpine 46–7, 80–1, 98–101
  herbaceous perennial 80–1, 98–101
Oxydendrum 126–9

Pachysandra 134–7
Pachystachys 130–1
Paeonia 86–7, 98–101
Pancratium 46–7, 90–1
Pandanus 102–3
Papaver
  alpine 46–7, 74–9, 98–101
  herbaceous perennial 74–9
Paradisea 46–7, 90–1
Parahebe 134–7
Parrotia 108–9
Parthenocissus 132–3, 144–5, 174–5
Passiflora 74–9, 130–1
Patrinia
  alpine 46–7
  herbaceous perennial 50–1, 98–101
Paulownia 66–9, 74–9
Peach 180–1, 184–5
Pear 114–15, 170–3, 180–1
Pelargonium 130–1
Pellaea 52–3, 98–101
Pellionia 130–1
Penstemon
  alpine 130–1
  herbaceous perennial 130–1
  shrub 134–7
Peperomia
  rosetted forms 158–9
  stem forms 130–1
Pernettya
  alpine 154–5
  shrub 118–19, 154–5
Petteria 62–3, 66–9
Petunia 48–9
Philadelphus 130–1, 132–3, 142–3
Philesia 154–5
Phillyrea 134–7
Philodendron 122–3, 130–1
Phlomis 150–1
Phlox
  alpine 46–7, 74–9, 98–101, 130–1
  herbaceous perennial 74–9
Phoenix 48–9
Phormium 98–101, 102–3
Photinia

deciduous 130–1, 132–3
  evergreen 134–7
Phygelius
  herbaceous perennial 98–101, 130–1
  shrub 130–1
Phyllodoce 118–19, 154–5
Phyllostachys 98–101
Physalis 98–101, 130–1
Physocarpus 138–41
Physostegia 98–101
Phyteuma 46–7
Picea
  species 66–9
  varieties 178–9
Pieris 118–19, 152–3
Pilea 130–1
Pinguicula 158–9
Pinus
  species 66–9
  varieties 178–9
Piptanthus 62–3, 66–9
Pittosporum
  indoor 130–1, 132–3
  outdoor 134–7
Plantago 46–7
Platanus 138–41
Platycerium 52–3
Platycodon
  alpine 46–7
  herbaceous perennial 50–1
Plectranthus 130–1
Plum
  rootstock 114–15, 132–3, 138–41
  varieties 170–3, 180–1, 184–5
Plumbago 130–1
Podocarpus
  species 66–9
  varieties 146–9
Polemonium 98–101
Polygala 46–7, 130–1
Polygonatum 86–7
Polygonum
  alpine 98–101, 130–1
  climber 138–41
  herbaceous perennial 98–101,
    130–1
Poncirus 66–9, 134–7
Populus 138–41
Potentilla
  alpine 46–7, 98–101, 102–3, 130–1
  herbaceous perennial 98–101, 102–3
  shrub 130–1, 132–3
Primula
  alpine 46–7, 74–9, 98–101, 158–9
  annual bedding plant 48–9
  herbaceous perennial
    bog 48–9, 98–101
    drumstick 74–9, 98–101
    primrose 98–101
Prunella 98–101
Prunus
  blackthorn 74–9, 98–101
  evergreen 134–7
  flowering cherry 174–5, 180–1, 184–5
  plum species 74–9

Pseudolarix 66–9
Pseudotsuga
  species 66–9
  varieties 178–9
Pteris 52–3, 98–101
Pterocarya 66–9
Ptilotrichum 130–1
Pulmonaria 98–101
Pulsatilla
  alpine 46–7, 74–9, 98–101
  herbaceous perennial 50–1, 98–101
Punica 48–9, 130–1
Puschkinia 46–7, 90–1, 98–101
Pyracantha 132–3, 134–7
Pyrethrum 98–101
Pyrola 98–101
Pyrus
  species 64–5, 66–9
  varieties 170–3, 180–1, 184–5

Quercus
  species 64–5, 66–9
  varieties and hybrids 176–7
Quince 170–3, 180–1, 184–5

Ramonda 46–7, 122–3
Ranunculus
  alpine 46–7, 98–101
  herbaceous perennial 50–1, 98–101
Raoulia 98–101
Raspberry 74–9
Red currant 142–3
Rhamnus 66–9, 132–3
Rhododendron
  azalea
    deciduous 108–9, 126–9, 176–7
    evergreen 132–3, 134–7
  hybrids
    large-leafed 108–9, 110–11,
      152–3, 176–7
    small-leafed 152–3
  species 40–1, 108–9, 110–11, 118–19
Rhodohypoxis 46–7, 98–101
Rhodotypos 138–41, 142–3
Rhoeo 130–1
Rhubarb 86–7
Rhus 74–9
Ribes 132–3, 138–41
Ricinus 48–9
Robinia
  species 62–3, 66–9, 74–9
  varieties 180–1, 184–5
Rodgersia 50–1, 98–101
Romneya 74–9
Romulea 46–7, 88–9
Rosa
  Floribunda or Hybrid Tea 138–9,
    182–3
  rootstock 64–5, 66–9, 142–3
  shrub 138–41, 176–7
  species 62–3, 66–9, 138–41
Rosmarinus 150–1
*Rubus 74–9, 112–13, 122–3, 130–1,
  132–3*
Rudbeckia 98–101

Ruellia 48–9, 130–1
Ruscus 134–7
Ruta 150–1

Sagina 98–101
Saintpaulia 48–9, 158–9
Salix
  alpine 130–1
  shrub or tree 138–41
Salvia
  herbaceous perennial 46–7, 98–
    101, 130–1
  shrub 150–1
Sambucus 138–41
Sanguinaria 46–7, 98–101
Sanguisorba 46–7
Sansevieria 86–7, 98–101, 166–7
Santolina 150–1
Saponaria
  alpine 46–7, 130–1
  herbacious perennial 98–101
Sarcococca 134–7
Sasa 98–101
Saxifraga
  alpine 46–7, 98–101, 130–1
  herbaceous perennial 98–101
  house plant 102–3
Scabiosa
  alpine 46–7, 130–1
  herbaceous perennial 46–7, 98–101
Schefflera 130–1
Schizanthus 48–9
Schizophragma 126–9
Schizostylis 86–7, 98–101
Schlumbergera 130–1
Sciadopitys 66–9
Scilla 46–7, 90–1, 96–7, 98–101, 166–7
Scindapsus 122–3, 130–1
Scrophularia 98–101
Scutellaria
  alpine 46–7, 130–1
  herbaceous perennial 50–1, 130–1
Seakale 74–9
Sedum
  alpine 46–7, 98–101, 130–1, 168–9
  herbaceous perennial 130–1
Selaginella 130–1
Sempervivum 102–3
Senecio
  alpine 46–7
  herbaceous perennial 50–1, 98–101
  house plant 130–1
  shrub 150–1
Sequoia 66–9, 146–9
Sequoiadendron 66–9, 146–9
Setcreasea 130–1
Shortia 98–101
Sidalcea 98–101
Silene 46–7
Sisyrinchium 98–101
Skimmia 134–7
Smilacina 86–7, 98–101
Smilax 86–7, 98–101
Smithiantha 48–9, 158–9
Solanum

bedding plant 48–9
house plant 48–9
shrub or tree 130–1
Soldanella 46–7, 98–101
Solidago 98–101
Sophora 66–9
Sorbaria 138–41
Sorbus 170–3, 174–5, 180–1, 184–5
Sparaxis 46–7, 88–9
Sparmannia 130–1
Spartium 68–9
Spathiphyllum 98–101
Spiraea 138–41
Sprekelia 90–1, 98–101
Stachys
    alpine 98–101
    herbaceous perennial 98–101
Stachyurus 106–9
Staphylea 106–9
Stephanotis 130–1
Sternbergia 90–1, 98–101
Stewartia 66–9, 106–9
Stokesia 98–101
Stranvaesia 106–9, 138–41
Strawberry 102–3
Strelitzia 86–7, 98–101
Streptocarpus
    indoor 160–1
    outdoor 48–9, 160–1
Streptosolen 130–1
Stromanthe 130–1
Styrax 64–5, 66–9, 106–9
Symphoricarpos 142–3
Symphyandra 46–7
Symphytum 50–1, 98–101
Syringa 98–101, 110–11, 114–15, 180–1

Tagetes 48–9
Tamarix 138–41
Tanacetum 98–101
Taxodium 66–9
Taxus 146–9
Tecophilaea 46–7, 90–1
Tellima 168–9
Teucrium
    alpine 132–3
    herbaceous perennial 98–101, 130–1
Thalictrum
    alpine 46–7, 98–101
    herbaceous perennial 98–101
Thlaspi 46–7
Thorn 174–5, 180–1
Thuja
    species 66–9
    varieties 146–9
Thujopsis 146–9
Thunbergia 48–9
Thymus 98–101, 130–1
Tiarella 98–101, 168–169
Tigridia 88–9
Tilia
    species 62–3, 64–5, 66–9
    varieties 180–1, 184–5
Tillandsia 98–101
Tolmiea 168–9

Torenia 48–9
Torreya 40–1, 66–9
Trachelospermum 130–1, 134–7
Tradescantia
    indoor 130–1
    outdoor 98–101
Tricyrtis 46–7
Trifolium 46–7
Trillium
    alpine 98–101
    rhizomatous 46–7, 86–7
Tritonia
    bedding plant 98–101
    bulbous 88–9
Trollius 98–101
Tropaeolum 40–1, 84–5
Tsuga
    species 66–9
    varieties 178–9
Tulipa 46–7, 90–1

Ulex
    species 66–9
    varieties 134–7
Ulmus
    species 66–9
    varieties 180–1, 184–5
Umbellularia 134–7
Uvularia 50–1, 98–101

Vaccinium
    alpine 154–5
    shrub or tree 118–19, 154–5
Valeriana 98–101
Vallota 90–1, 166–7
Veitchberry 112–13, 122–3, 130–1
× *Venidio–Arctotis* 98–101, 130–1
Venidium 50–1, 98–101, 130–1
Verbascum
    alpine 46–7, 74–9, 98–101, 130–1

herbaceous perennial 74–9
Verbena 98–101, 130–1
Veronica
    alpine 98–101, 130–1
    herbaceous perennial 98–101
Viburnum
    evergreen summer-flowering 134–7
    species 62–3, 64–5, 66–9
    winter-flowering 138–41

Vinca
    alpine 150–1
    shrub 134–7, 150–1
Viola
    alpine 46–7, 98–101
    herbaceous perennial 98–101
    pansy 48–9, 98–101
Vitex 66–9
Vitis 122–3, 130–1, 144–5
Vriesia 98–101

Wahlenbergia 46–7, 98–101, 130–1
Walnut 66–9, 106–9
Watsonia 88–9
Weigela 132–3, 142–3
White currant 142–3
Wisteria 174–5
Wulfenia 46–7

Yucca 98–101, 102–3

Zantedeschia 86–7, 98–101
Zauschneria 130–1
Zebrina 130–1
Zelkova 130–1, 66–9
Zenobia 118–19, 134–7
Zephyranthes 46–7, 90–1
Zinnia 48–9
Zygocactus 130–1

Acknowledgements

The author and publishers wish to extend their thanks to the following individuals, organizations, and institutions who have given invaluable help and advice during the preparation of this book: Blagg & Johnson Ltd, Newark; Burton McCall (Horticultural) Ltd., Leicester; Humex Ltd, Weybridge; Jiffy Garden Products, Rotherfield; Margaret Lee-Elliott; Ministry of Agriculture, Fisheries and Food (Entomology Department), Cambridge; and George Ward (Moxley) Ltd, Darlaston.

Artists: Linda Broad, Pamela Dowson, Chris Forsey, William Giles, Tony Graham, Vana Haggerty, Terry Lawler, Colin Salmon, Mike Saunders and Ralph Stobart.
Picture research: Mari Zipes.

All artwork in this book has been based on photographs specially commissioned from R. Robinson of the Harry Smith Horticultural Photographic Collection, except for those indicated below: p. 42 (rhizome, corm and bulb), p. 45 (corm cross-section), and p. 46 (scaly bulb, tunicate bulb) which are all based on diagrams in *Plant Propagation: Principles and Practices* by Hudson T. Hartmann and Dale E. Kester (Prentice-Hall: New Jersey, 2nd edn 1968). Reprinted by permission of Prentice-Hall Inc., New Jersey, USA.

Typesetting by SX Composing Ltd, Raleigh, Essex
Origination by M&E Reproductions, North Fambridge, Essex

THE R.H.S. ENCYCLOPEDIA OF PRACTICAL GARDENING

EDITOR-IN-CHIEF: CHRISTOPHER BRICKELL

A complete range of titles in this series is available from all good bookshops or by mail order direct from the publisher. Payment can be made by credit card or cheque/postal order in the following ways:
BY PHONE   Phone through your order on our special CREDIT CARD HOTLINE on 01933 443863; speak to our customer service team during office hours (9am to 5pm) or leave a message on the answer machine, quoting your full credit card number plus expiry date and your full name and address.
BY POST   Simply fill out the order form below (it can be photocopied) and send it with your payment to: MITCHELL BEAZLEY DIRECT, 27 SANDERS ROAD, WELLINGBOROUGH, NORTHANTS, NN8 4NL.

ISBN	TITLE	PRICE	QUANTITY	TOTAL
1 84000 160 7	GARDEN PLANNING	£8.99		
1 84000 159 3	WATER GARDENING	£8.99		
1 84000 157 7	GARDEN STRUCTURES	£8.99		
1 84000 151 8	PRUNING	£8.99		
1 84000 156 9	PLANT PROPAGATION	£8.99		
1 84000 153 4	GROWING FRUIT	£8.99		
1 84000 152 6	GROWING VEGETABLES	£8.99		
1 84000 154 2	GROWING UNDER GLASS	£8.99		
1 84000 158 5	ORGANIC GARDENING	£8.99		
1 84000 155 0	GARDEN PESTS AND DISEASES	£8.99		
			POSTAGE & PACKING	£2.50
			GRAND TOTAL	

Name ....(BLOCK CAPITALS)

Address ....

....Postcode..............................

I enclose a cheque/postal order for £.................made payable to Octopus Publishing Group Ltd. or:

Please debit my: Access ☐ Visa ☐ AmEx ☐ Diners ☐ account

by £............................ Expiry date ....

Account no ☐☐☐☐☐☐☐☐☐☐☐☐☐☐☐☐☐☐

Signature ....

Whilst every effort is made to keep our prices low, the publisher reserves the right to increase the price at short notice.
Your order will be dispatched within 28 days, subject to availability. Free postage and packing offer applies to UK only. Please call 01933 443863 for details of export postage and packing charges

Registered office: 2-4 Heron Quays, London E14 4JP Registered in England no 3597451
THIS FORM MAY BE PHOTOCOPIED.